Physical science for technicians

Physical science for technicians

I. McDonagh
Senior lecturer in engineering,
Wirral Metropolitan College

and

G. Waterworth
Lecturer in electrical engineering,
Leeds Polytechnic

illustrated by
R. P. Phillips
Lecturer in mechanical engineering,
Wirral Metropolitan College

Edward Arnold

First published in Great Britain 1986
by Edward Arnold (Publishers) Ltd
41 Bedford Square
London WC1B 3DQ

Edward Arnold (Australia) Pty Ltd
80 Waverley Road
Caulfield East
Victoria 3145
Australia

British Library Cataloguing in Publication Data

McDonagh, I.
Physical science for technicians.
1. Science
I. Title II. Waterworth, G. III. Phillips, R. P.
500.2′0246 Q160.2

ISBN 0-7131-3510-7

Text set in 10/11 pt IBM Press Roman by TEC SET, Sutton, Surrey.
Printed in Great Britain by Richard Clay (The Chaucer Press) Ltd, Bungay, Suffolk.

Contents

Preface

This book has been written to meet the requirements of the Business & Technician Education Council (BTEC) standard unit Physical science I (U80/682), which is a core unit in a wide range of National Certificate and Diploma programmes in engineering and construction and also in the Technician Studies scheme for TVEI and other pre-vocational students.

The aim of the book is to help the student to understand the fundamental physical-science concepts which form a common base for further study of both science and technology.

Laboratory work is an essential part of the BTEC unit, and throughout the book we have described experiments, some of which can be done at home while others require the facilities provided in a physical-science laboratory.

SI units have been used throughout.

I. McDonagh
G. Waterworth

Acknowledgements

The authors and publishers would like to thank the following for permission to reproduce photographs: R. A. Davenport (page 11), Metal Treatments (Birmingham) Ltd (page 12), Austin Rover Group (page 13, top), Zinc Development Association (page 13, bottom), Norbar Torque Tools Ltd (page 29), Tangye Ltd (page 42), Lucas Kienzle Instruments Ltd (page 62), Ministry of Defence (page 65), Marconi Radar Systems Ltd (page 74), Philips Medical Systems (page 75), *Soldier* magazine (page 79), John Holden (Ocean Publishing and Technical) (page 84, top), Foto Pietro Magni, Milan, and Impregio & Co. Ltd (Ghana) (page 84, bottom), Felco Hoists Ltd (page 87), Maurer (UK) Ltd (pages 100 and 101), Sinclair Vehicles Ltd (page 105, top), Ford Motor Company Ltd (page 105, bottom), Lucas Electrical Ltd (page 106), Central Electricity Generating Board (page 107), Bach-Simpson (UK) Ltd (pages 113 and 161), Fenwal Electronics and Electrautom Ltd (page 128), Ferranti Measurements Ltd (page 135), Edward Wilcox & Co. Ltd (page 137), British Steel Corporation (page 153), Coats Pacesetter Ltd (page 155), GEC Power Transformers Ltd (page 156), Inductotherm (Europe) Ltd (page 157).

1 Physical quantities

1.1 The International System of Units

The international system of units, called SI units, was established by international agreement in 1960 and consists of seven base units plus two supplementary units as shown in Table 1.1

Table 1.1 Base and supplementary SI units (the base units are shown in bold type)

Quantity	Unit	Symbol
Length	**metre**	m
Mass	**kilogram**	kg
Time	**second**	s
Electric current	**ampere**	A
Temperature	**kelvin**	K
Luminous intensity	**candela**	cd
Amount of substance	**mole**	mol
Plane angle	radian	rad
Solid angle	steradian	sr

1.2 Derived units

From the base units shown in Table 1.1, all other units used in physical science are derived. Table 1.2 lists the quantities used in this book.

Table 1.2 Derived SI units

Quantity	Symbol	Base SI units	Derived unit
Area	A	$\mathrm{m} \times \mathrm{m}$	m^2
Volume	V	$\mathrm{m} \times \mathrm{m} \times \mathrm{m}$	m^3
Density	ρ	kg per m^3	$\mathrm{kg/m^3}$
Force and weight	F, W	$\mathrm{kg\,m/s^2}$	N (newton)
Pressure	P	$\mathrm{kg/m\,s^2}$	$\mathrm{N/m^2}$ = Pa (pascal)
Energy	E or U	$\mathrm{kg\,m^2/s^2}$	N m = J (joule)
Power	P	$\mathrm{kg\,m^2/s^3}$	J/s = W (watt)
Frequency	f	$\mathrm{s^{-1}}$	Hz (hertz)
Electric charge	Q	A s	C (coulomb)
Electrical potential difference	V	$\mathrm{kg\,m^2/(s^3\,A)}$	V (volt)
Electrical resistance	R	$\mathrm{kg\,m^2/(s^3\,A^2)}$	Ω (ohm)

The names of some of the base and derived SI units have been chosen in recognition of the work done by famous scientists in the past; for example, the unit of force is the newton and has been named after Sir Isaac Newton.

1.3 Prefixes for SI units

The unit of length, the metre, is too large when it is required to measure say the diameter of a sewing needle (typically 0.001 m) and too small when measuring say the distance between Land's End and John O'Groats (approximately 1 380 000 m). For this reason, prefixes or multipliers are used so that a quantity can be expressed in realistic terms. Using prefixes, the diameter of the sewing needle is 1 mm and the distance between Land's End and John O'Groats is approximately 1380 km. Table 1.3 shows the preferred prefixes in bold type.

Table 1.3 Prefixes for use with SI units (the preferred prefixes are shown in bold type)

Prefix	Symbol	Multiply by
tera	T	10^{12} = 1 000 000 000 000
giga	G	10^{9} = 1 000 000 000
mega	M	10^{6} = 1 000 000
kilo	k	10^{3} = 1 000
hecto	h	10^{2} = 100
deca	da	10^{1} = 10
deci	d	10^{-1} = 0.1
centi	c	10^{-2} = 0.01
milli	m	10^{-3} = 0.001
micro	μ	10^{-6} = 0.000 001
nano	n	10^{-9} = 0.000 000 001
pico	p	10^{-12} = 0.000 000 000 001
femto	f	10^{-15} = 0.000 000 000 000 001
atto	a	10^{-18} = 0.000 000 000 000 000 001

Examples of use of prefixes

$$4\,000\,000\,\text{W} = 4 \times 10^{6}\,\text{W} = 4\,\text{MW}$$

$$0.000\,000\,003\,\text{m} = 3 \times 10^{-9}\,\text{m} = 3\,\text{nm}$$

$$200\,000\,000\,\text{N/m}^2 = 200 \times 10^{9}\,\text{N/m}^2 = 200\,\text{GN/m}^2$$

Care must be taken when using prefixes with the unit for mass, the kilogram. For example

$$60\,000\,\text{kg} = 60 \times 10^{3}\,\text{kg} = 60\,\text{Mg} \quad (\textit{not}\ 60\,\text{kkg})$$

This anomaly is because the kilogram has been chosen in preference to the gram as the unit for mass, since for practical purposes the gram was considered to be too small.

When using prefixes, care must be taken when a unit is raised to a power, e.g. is squared or cubed, as the power applies to *both* the unit *and* the prefix as illustrated in the following examples.

Example Express the following in SI base units: (a) 25 km squared, (b) 4 mm cubed, (c) the square root of $1600\,\text{mm}^2$.

a) $(25\,\text{km})^2 = (25 \times 10^3\,\text{m})^2$

$= 25^2 \times 10^{3\times 2}\,\text{m}^2$

$= 625 \times 10^6\,\text{m}^2$

i.e. 25 km squared $= 625 \times 10^6\ \text{m}^2$

Notice how the multiplier has been squared by multiplying the index (i.e. 3) by 2.

b) $(4\,\text{mm})^3 = (4 \times 10^{-3}\,\text{m})^3$

$= 4^3 \times 10^{-3\times 3}\,\text{m}^3$

$= 64 \times 10^{-9}\,\text{m}^3$

i.e. 4 mm cubed $= 64 \times 10^{-9}\,\text{m}^3$

Notice how the multiplier has been cubed by multiplying the index (i.e. –3) by 3.

c) $\sqrt{(1600\,\text{mm}^2)} = (1600 \times 10^{-6}\,\text{m}^2)^{1/2}$

$= 1600^{1/2} \times 10^{-6\times 1/2}\,\text{m}^{2\times 1/2}$

$= 40 \times 10^{-3}\,\text{m}$

i.e. the square root of $1600\,\text{mm}^2 = 40 \times 10^{-3}\ \text{m}$

Exercises on chapter 1

1 Using preferred prefixes, rewrite the following quantities: (a) 63 000 Pa, (b) 0.000 000 05 F, (c) 487 000 000 kg, (d) 0.027 54 m.
2 Given that 150 MN = 150×10^6 N, express the following in a similar form: (a) 92.5 pF, (b) 7568 km, (c) 56.3 Tm, (d) 760 Gg.
3 Express the following in SI base units: (a) $(12.5\ \text{mm})^4$, (b) $\sqrt{(49\,\text{kW}^2)}$, (c) $\sqrt[3]{(125\ \mu\text{m}^3)}$, (d) $(20\,\text{M}\Omega)^3$.

2 Chemical reactions

2.1 Introduction

Some of the topics in this book - such as oxidation, rusting, electric cells, electrolysis, and electroplating - will require a knowledge of atomic structure and of chemical processes. The aim of this chapter is to provide this necessary background and also to describe, in simple terms, the chemical processes involved in burning and rusting and to discuss them as examples of chemical reactions.

2.2 Elements and compounds

All materials are made up of atoms. There are 92 different types of atom which occur naturally. A substance which consists of only one type of atom is called an *element*.

Elements can combine with other elements to form *compounds* in which the atoms of one element are joined to those of another. For example, when sodium atoms (chemical symbol Na) and chlorine atoms (chemical symbol Cl) combine, the result is sodium chloride (chemical symbol NaCl) or common salt. Hydrogen and oxygen combine to form water (H_2O); carbon and oxygen combine to form carbon dioxide (CO_2).

Most atoms have a strong tendency to link up with other different atoms, so most substances in the natural world are compounds.

2.3 The world of atoms

Atoms are the smallest particles of elements which can take part in chemical reactions, but they are themselves made up of even smaller more fundamental particles of many different types.

One very useful model of the atom was suggested by the Danish physicist Niels Bohr (1885-1962). This considers all atoms as being made up of a central *nucleus* with surrounding *electrons* which rotate around the nucleus rather like the planets rotate around the sun.

The electrons are arranged in rings (or more precisely in shells) around the nucleus and at different distances from it. (A better model is obtained if we consider the electrons more as puffs of smoke surrounding the nucleus.)

Electrons are extremely small and light particles, and each carries a single negative electric charge. In comparison, the central nucleus is much bigger and heavier. It consists of particles called protons and neutrons which are tightly packed together. A proton carries a positive charge equal and opposite to the negative charge on an electron, while neutrons carry no charge. Since in an atom there are as many protons as there are electrons, then the central nucleus carries a positive charge equal and opposite to the total negative charge on all the surrounding electrons. The atom is thus electrically neutral.

Figure 2.1 shows a Bohr atomic model of the element hydrogen (chemical symbol H) which has a central nucleus and a single orbiting electron.

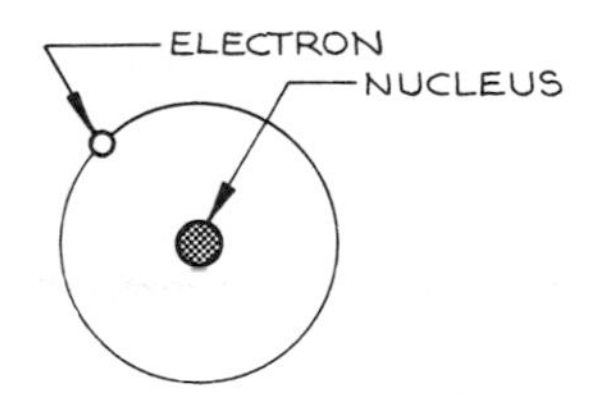

Fig. 2.1 The hydrogen atom

Zinc (chemical symbol Zn) has 30 electrons in the surrounding shells as shown in fig. 2.2. Notice that the electrons do not all reside in one single shell, since there is a maximum number of electrons that can exist in any particular shell. This maximum is different for different shells.

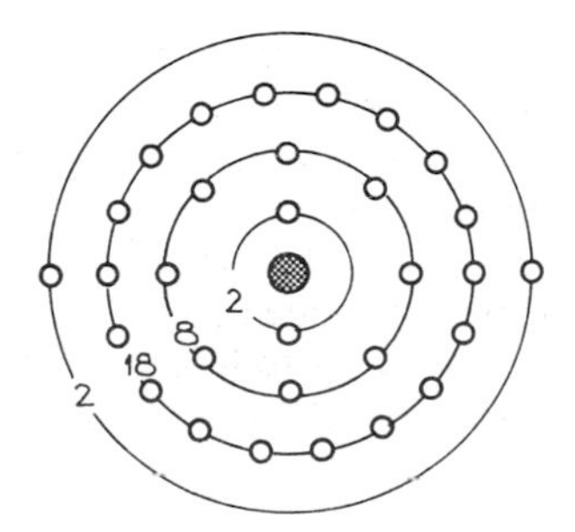

Fig. 2.2 The zinc atom

The first shell can only hold up to two electrons. The second shell can hold up to eight electrons. The third shell can hold up to eighteen electrons but in many elements it is stable with eight. Elements with full outer shells tend to be very stable (unreactive), while elements with almost empty or almost full outer shells tend to be very unstable (reactive). Table 2.1 shows the number of outer-shell electrons for various elements.

2.4 Chemical reactions

A chemical reaction is an interaction between two or more substances involving a rearrangement of the atoms.

Atoms combine together to form *molecules*.

Table 2.1 Some elements and their number of outer-shell electrons

Element	Chemical symbol	Number of outer-shell electrons
Aluminium	Al	3
Calcium	Ca	2
Carbon	C	4
Chlorine	Cl	7
Chromium	Cr	7
Copper	Cu	1
Fluorine	F	7
Helium	He	2
Hydrogen	H	1
Iodine	I	7
Iron	Fe	8
Lead	Pb	4
Magnesium	Mg	2
Mercury	Hg	2
Neon	Ne	8
Nickel	Ni	8
Nitrogen	N	5
Oxygen	O	6
Phosphorus	P	5
Platinum	Pt	8
Potassium	K	1
Silicon	Si	4
Silver	Ag	1
Sodium	N	1
Sulphur	S	6
Tin	Sn	4

Atoms of a single element may combine to form molecules of that element - for example, hydrogen atoms (H) combine together to form hydrogen molecules (H_2); chlorine atoms combine to form chlorine molecules (Cl_2). They can also combine with atoms of different elements to form compound molecules.

There are a number of possible ways in which atoms bond together. We shall consider two.

First, let us consider the molecular structure of sodium chloride (common salt), which is a compound of sodium (Na) and chlorine (Cl). The structures of the sodium and chlorine atoms are shown in fig. 2.3. Notice that the sodium atom has only one electron in its outer shell, while the chlorine atom has seven. The chlorine atom thus lacks one electron for a complete stable outer shell. When a sodium and a chlorine atom come together, the sodium atom loses its outer electron to the chlorine atom. Having lost one electron, the sodium atom is then referred to as a sodium ion (and has a net positive charge). The chlorine atom which has gained one electron is then referred to as a chloride ion (and has a net negative charge).

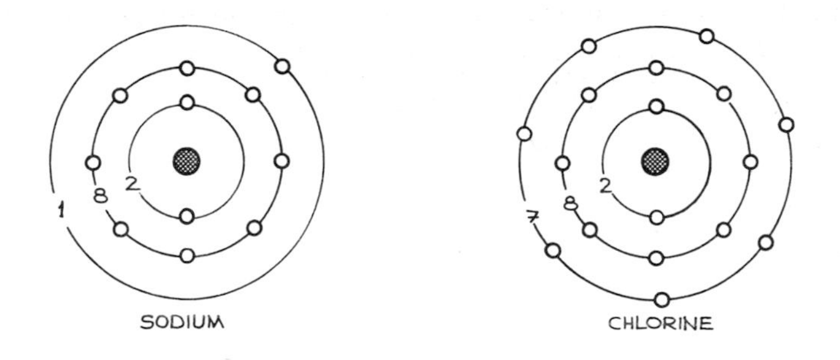

Fig. 2.3 The structure of sodium and chlorine atoms

Both the ions formed have stable electron arrangements and, since they have opposite charges, they are held together by strong electrostatic forces of attraction to form a sodium-chloride molecule. This type of bonding is known as *ionic bonding*.

Another type of bonding - known as *covalent bonding* - involves the *sharing* of electrons rather than a complete transfer. Consider the chlorine molecule, shown in fig. 2.4 - this consists of two chlorine atoms bonded together. One electron from each atom is donated to form a pair of electrons which is shared between both atoms and holds them together.

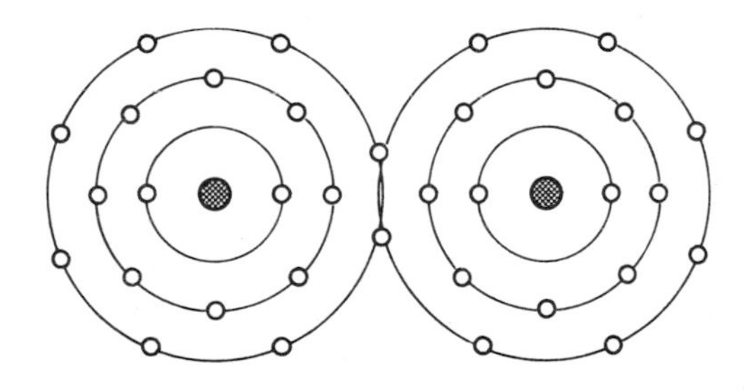

Fig. 2.4 The chlorine molecule

2.5 Air

Air is a mixture of gases and consists approximately of $\frac{4}{5}$ nitrogen and $\frac{1}{5}$ oxygen. There is also a small proportion of carbon dioxide, and even smaller proportions of the rare gases argon, neon, helium, krypton, and xenon. Water vapour is also present in small quantities, the exact proportion depending on the air temperature.

The composition of air remains fairly constant. Although oxygen is continually being extracted from it by living organisms, by combustion, and by slow oxidation, this is balanced by plants taking in carbon dioxide via their leaves and using sunlight to produce free oxygen in a process called photosynthesis. The absorption of oxygen by animals is in turn accompanied by the release of carbon dioxide.

2.6 Oxidation

Oxidation is the process of combining oxygen with another substance or of removing hydrogen from it.

The compounds formed when elements react with oxygen are referred to as *oxides*.

Metals form oxides when exposed to air, e.g. aluminium oxide or iron oxide (rust). This process is referred to as *corrosion*. When materials such as wood or paper burn in air, oxides are again formed. This process is referred to as *combustion* (or burning). Rusting and combustion are both processes of oxidation. In each case, it is a chemical reaction which is taking place.

The combining of a metal with oxygen results in a compound which has a greater mass than the original metal. This can be demonstrated by a simple experiment.

Experiment 2.1 To show the effect of combining oxygen with magnesium.

Place a piece of magnesium ribbon in a crucible and weigh them together with the crucible lid.

Now heat the crucible with a bunsen burner as shown in fig. 2.5, ensuring that the lid is kept in place so that the fumes do not escape.

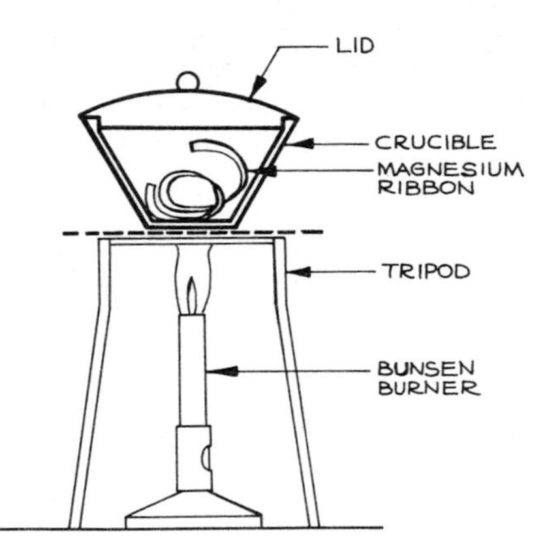

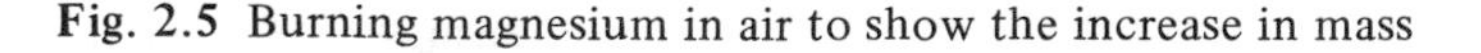

Fig. 2.5 Burning magnesium in air to show the increase in mass

When all of the magnesium has burned, allow the crucible to cool and then reweigh it. It will be observed that an increase in mass has taken place – this is due to the metal combining with oxygen in the air to form magnesium oxide.

2.7 Corrosion

The term corrosion is used to describe the slow reaction which occurs when metals are in contact with gases in the atmosphere or with liquids.

The corrosive environment may be a chemical such as an acid or another fluid capable of conducting an electric current, such as salt water or moist polluted air. Such a fluid is called an *electrolyte*, and the process is an example of *electrochemical corrosion* (see chapter 12). In either case the result is the eventual conversion of the metal surface into a crumbly compound formed by the reaction between the metal and the corrosive fluid.

For example, when a steel or an iron object such as a garden tool is left outside in the rain for several days, the result is that it goes rusty. Rusting is an example of electrochemical corrosion.

Corrosion is much faster when two dissimilar metals are in contact in the presence of a conducting liquid such as rain water. The combination forms a *corrosion cell* very similar to the electric cell of a normal torch battery (see chapter 11).

The basic requirements of a corrosion cell are a conducting liquid (the electrolyte) and two dissimilar metals (the electrodes) which must be in contact so that an electric current can flow. One of the metals acts as the positive electrode (the anode) and the other metal acts as the negative electrode (the cathode).

Which metal acts as the anode and which as the cathode is determined by their relative *electropotential*. The metal with the *higher* electropotential is always the *cathode*, while the metal with the *lower* electropotential is always the *anode*. It is always the anode which corrodes away. Table 2.2 lists commonly used metals (and carbon, which behaves like a metal in this respect) in *descending* order of electropotential.

Referring to Table 2.2, silver is the metal with the highest electropotential and thus will always be the cathode if paired with any other metal in the list, while magnesium has the lowest electropotential and will always be the anode

Table 2.2 Commonly used elements in descending order of electropotential

Carbon
Silver
Stainless-steel
Bronze
Copper
Lead
Tin
Nickel
Cast-iron
Wrought-iron
Mild steel
Chromium
Zinc
Aluminium
Magnesium

if paired with any other metal in the list - i.e. the magnesium will always corrode, while the silver will not.

If two elements from this list are in contact, then it is always the one lower in the list which acts as the anode and therefore corrodes.

Steel is best coupled with metals such as cadmium, aluminium, zinc, and magnesium. Coupling with copper, say, will result in the steel acting as the anode and corroding.

Example 1 Why is it acceptable to use copper rivets with large steel sections but never steel rivets with large copper sections?

In either case the steel will act as the anode and will tend to corrode when electrochemical action occurs. This is obviously dangerous in the case where the rivets are made of steel and corrode.

Example 2 Why is it appropriate to zinc-plate sheet-steel pressings to prevent corrosion of the steel?

When electrochemical action occurs, the zinc acts as the anode and discourages corrosion of the steel.

2.8 Rusting

A piece of iron left out in the rain will go rusty even though it is apparently not in contact with another metal. How then does electrochemical corrosion take place?

The corrosion of iron by rusting is a complex process, but it is the presence of impurities in the iron which allows the electrochemical action to take place. The iron and the impurities form the electrodes of tiny cells with water as the electrolyte (see chapter 12), and localised currents flow through the water between these electrodes and complete their path through the body of the iron.

No rusting can take place in the absence of moisture. (Motor cars hardly rust at all in hot dry areas like North Africa.) Pure water is not a very good electrolyte, but dissolved salts or dissolved carbon dioxide from the air greatly increase the rate of rusting.

The requirements for rusting are thus liquid water and oxygen.

Iron or steel immersed in water that has been boiled long enough to expel the dissolved air, will not rust, since the dissolved gases are required to enable the water to act as an electrolyte. Also at ordinary temperatures moist air does not cause rusting (e.g. a penknife kept in the pocket will not rust).

Since rust on the surface of iron or steel is porous and also flakes off, it does not form a protective surface on the layer of iron beneath it as is the case with the oxides formed on the surface of aluminium or copper, say. The rust area thus gets bigger, and eventually the whole of the iron or steel is converted into rust.

Other examples of corrosion of iron or steel of industrial importance are the corrosion of metal water tanks at the water line and the corrosion of buried metals.

Corrosion of the sill and lower body shell of a van, due to rusting

2.9 Corrosion prevention and metal protection

Most corrosion is due to electrochemical action, and some means of protecting metals against contact with an electrolyte is the best method of prevention.

Some metals - such as copper - corrode to form a thin film of oxide or carbonate which then protects them from further corrosion. Use is made of this effect in ***stainless steel*** by the addition to the steel of a small percentage of chromium. A thin film of chromium oxide of atomic dimensions forms over the surface of the steel and protects it from rusting. This does not happen with ordinary iron or steel since the rust is flakey and porous and does not form a protective layer.

Metal surfaces can be protected from corrosion by the application of a protective covering. The protective methods include plating with zinc (galvanising), tin, nickel, chromium, or silver, all of which resist atmospheric attack.

Anti-corrosion paints are also widely used, particularly to protect iron and steel from sea water.

Two modern methods of paint application are *cathodic electropainting* and *powder coating*.

Electropaint consists of a resin and a pigment treated with alkali and soluble in water. The paint is contained in a tank which is connected to the negative rail of a d.c. supply. The metal articles to be coated are connected to the positive rail of the supply and are passed through the paint so that the charged resin and pigment particles attach themselves to the articles by the action of the electric current.

In *powder coating*, electrostatically charged powder is sprayed on to the metal to be coated, which is then cured in an oven at 280°C.

Both of these methods provide better paint adhesion.

Phosphoric acid and zinc and manganese phosphates are all used in metal-finishing. These solutions form a crystalline phosphate coating which improves paint adhesion and prevents rust spreading beneath the paint film.

Chromium, as well as being added to steel to render it stainless, is also used to plate metals (chrome-plating). The plating is done electrolytically, as discussed in chapter 12.

Zinc is used as a coating on iron and mild steel to protect them from rusting. The iron is first 'pickled' in hydrochloric or sulphuric acid and sand to remove grease and surface impurities. It is then immersed in a bath of molten zinc, the thickness of the resulting coating depending on the temperature of the molten metal. Iron thus coated with zinc is referred to as 'galvanised iron'.

Where two dissimilar metals have to be used together, electrochemical corrosion can be prevented by separating the two metals using a non-metallic insulator such as a piece of plastic etc., thus preventing current flow.

Example Why is nickel-chrome wire a more suitable material than steel wire for use in electric fires?

One reason is that steel corrodes much more easily than nickel-chrome. Additionally, the nickel-chrome wire has a higher electrical resistance than steel.

De-rusting steel by immersion in tanks containing hydrochloric and phosphoric acids, for the rapid removal of rust before phosphating and painting.

A Rover body shell entering an electrocoating dip at Austin Rover, Cowley, where the latest cathodic process is used for paint application

A steel girder being immersed in a galvanising tank to provide a protective coating of zinc

Exercises on chapter 2

1 (a) Name the two main constituent gases in air. (b) Name the element which always constitutes part of an oxide.

2 When powdered copper is heated strongly in air for several minutes and is then allowed to cool, it turns black and its mass increases. (a) What elements are involved in the reaction? (b) What is the type of compound formed by the reaction?

3 A piece of iron is placed in each of the following four conditions: (a) in dry air, (b) in an atmosphere of air and steam produced by boiling water, (c) in a beaker containing tap water, (d) in a beaker containing distilled water. For each case, state whether rusting will occur and why.

4 (a) State what is meant by the term *chemical reaction*. (b) Give an example of a chemical reaction.

5 State what are meant by the following: (a) oxidation, (b) oxide, (c) element, (d) compound.

6 Complete the following statements:

a) Elements can combine with other elements to form ______________.
b) Atoms combine together to form ______________.
c) Hydrogen and oxygen combine to form ______________.
d) The Bohr atomic model of the atom consists of a central nucleus of particles surrounded by ______________.
e) Electrons carry a ______________ charge.
f) Common salt is a compound of ______________ and ______________.
g) The compounds formed when elements react with oxygen are referred to as ______________.

7 State the basic requirements of a corrosion cell.

8 State the basic requirements for electrochemical corrosion. Explain how the following methods are used to reduce corrosion: (a) painting, (b) chrome-plating.

9 In the following cell arrangements, state the electrode which will act as the anode: (a) zinc-carbon, (b) nickel-iron, (c) silver-sodium, (d) carbon-lithium.

10 Iron nails are put into three test tubes and left for several days. Tube (a) contains dry air and no water. Tube (b) contains water and no air. Tube (c) contains air and water. State in which cases rusting will take place.

11 State four methods that are used to protect iron and steel from rusting.

12 Complete the following statements:

a) Carbon burns in air or oxygen to form ______________.
b) When copper is heated in air or oxygen, the black surface coating produced is called ______________ ______________.
c) When magnesium is burned in air, the overall weight ______________.

13 When atoms bond together to form molecules, in one form of bonding a complete transfer of electrons takes place, and in another form the electrons are shared. Name the two forms of bonding and give an example of each.

14 State why a zinc or aluminium plate is often bolted on to the bottom of steel boats, and explain why the zinc or aluminium is often referred to as a 'sacrificial anode'.

3 Hooke's law

3.1 Definitions

If a rubber band is pulled, its length will increase. The pulling or stretching force on the band is known as a *tensile* force, and the *increase* in length of the band is termed the *extension* - i.e. a tensile or stretching force will produce an extension (see fig. 3.1(a)).

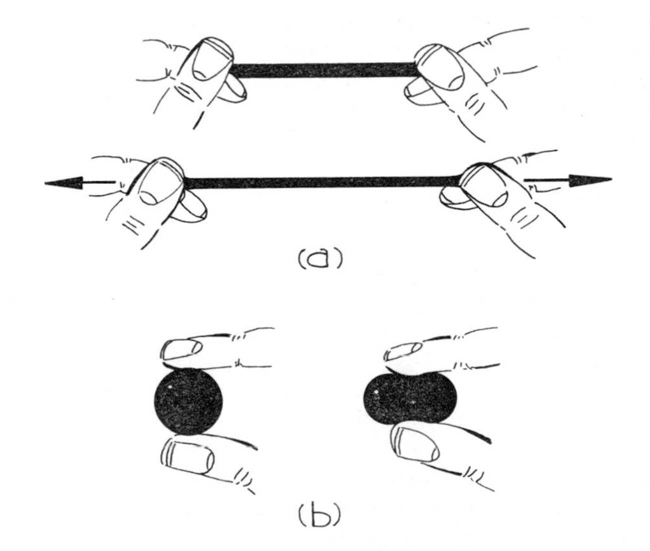

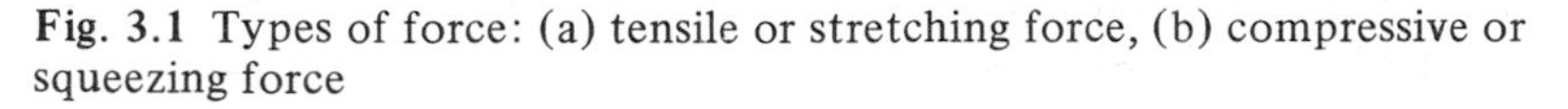
Fig. 3.1 Types of force: (a) tensile or stretching force, (b) compressive or squeezing force

When a rubber ball is squeezed, the diameter of the ball in the direction of the squeezing force is reduced. The squeezing force on the ball is known as a *compressive* force, and the decrease in diameter is termed the *compression* - i.e. a squeezing or compressive force will produce a *compression* (see fig. 3.1(b)).

If materials subjected to external tensile or compressive forces return to their original size or shape upon removal of the forces, they are known as *elastic* materials. Materials which remain permanently distorted upon removal of external forces are known as *plastic* materials. The effect of external forces on elastic materials only will be considered here.

3.2 Relationship between applied force and extension

The relationship between applied force and extension in elastic materials can be determined experimentally. Described here are two experiments, the first of which can be done at home on the kitchen table, while the second would normally be carried out in the physical-science or engineering-science laboratory.

Experiment 3.1 To find the relationship between applied force and the extension of a rubber band.

For this experiment, select a rubber band with a circumference of approximately 250 mm and made from material not more than 3 mm wide. Suspend the band from a steel knitting needle or a pencil which is supported between two books or two bottles of equal height as shown in fig. 3.2. Attach a simple hook made from a soft-wire paper-clip to the lower end of the rubber band. Forces may be applied to the rubber band in equal steps or increments by hanging steel washers, all having the same mass, on to the paper-clip hook. Alternatively, a simple sling may be attached to the hook into which either 10p or 2p coins may be placed.

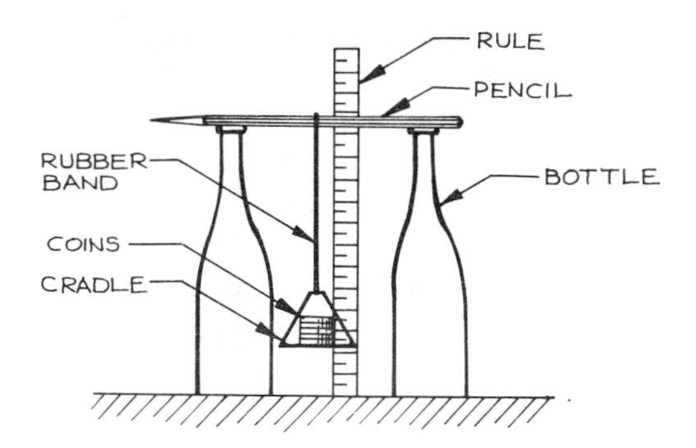

Fig. 3.2 Experiment on an elastic band

With the hook in place, sufficient washers or coins are added to take up any slack in the rubber band - i.e. the band is slightly tensioned. With the table top as a datum (i.e. a fixed point from which to make measurements), the distance to the lower end of the rubber band is measured with a rule and the measurement is recorded. A single washer or coin is added to the hook, thus applying an increment of force, and the distance from the datum to the lower end of the band is remeasured and recorded. This should be repeated about eight times.

The results from such an experiment are shown in Table 3.1, from which the graph in fig. 3.3 has been drawn. (It should be noted that in mathematics it is usual to plot the controlled variable, i.e. the number of coins or washers in this case, along the horizontal or x-axis. However, in science, where force

Table 3.1 Force–extension results for a rubber band

Number of objects	Distance from table top (mm)	Extension (mm)
0	167	0
1	165	2
2	163	4
3	161	6
4	159	8
5	157	10
6	155	12
7	153	14
8	151	16

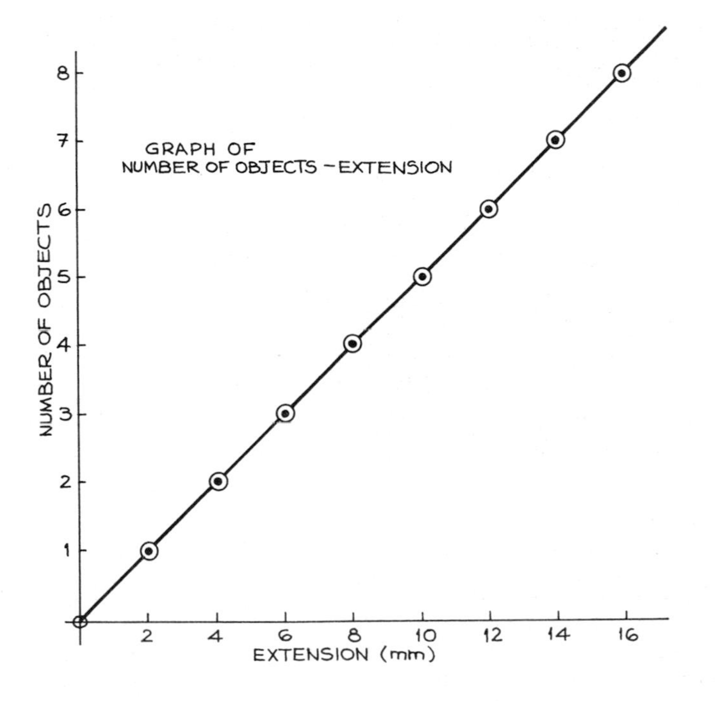

Fig. 3.3 Force–extension graph for an elastic band

or torque or moment is plotted against displacement, it is usual to plot the *displacement* along the horizontal axis.)

The graph in fig. 3.3 is a straight line. When a straight line is obtained from a set of readings, it indicates that for each increase of, say, 2 N in force there will be a corresponding extension of, say, 1 mm;

i.e. a force of 2 N produces an extension of 1 mm,
a force of 4 N produces an extension of 2 mm,
a force of 6 N produces an extension of 3 mm,

and so on.

In such a case, we say that the extension is *directly proportional* to the applied force. Thus, the conclusion for this experiment is that 'Within the range of forces used, the extension is directly proportional to the applied force.'

In the following experiment, which is presented as a formal report, the information enclosed in square brackets, [. . .], would not appear in the actual report.

Experiment 3.2

Object
To find the relationship between applied force and the extension of steel wire.

Apparatus
Hooke's-law apparatus, steel wire, masses, micrometer, and metre rule.
[The Hooke's-law apparatus shown in fig. 3.4(a) consists of a wall bracket from which is suspended the steel wire. Attached to the lower end of the wire is a vernier scale which is located in a fixed vertical guide. The purpose of the vernier scale is to divide each of the 1 mm divisions on the fixed scale into 10 equal parts so that the extension of the wire can be measured in increments or steps of 0.1 mm. On the lower end of the vernier scale, there is a hook to which a load-hanger may be attached.]

Method
The apparatus was set up as shown in fig. 3.4(a) and the diameter and length of the wire were measured. A small mass was added to the load-hanger to take up the slack in the steel wire, and the initial vernier-scale reading was noted. [The initial reading on the vernier scale is shown in fig. 3.4(b). Referring to fig. 3.4(b), the scale reading is 8.4 mm, since zero on the vernier scale lies between 8 mm and 9 mm on the fixed scale and the 'point of coincidence' where a graduation on the vernier is in line with a graduation on the fixed scale (indicated by the arrow) is at point 4 on the vernier – i.e. the reading is 8 mm + 0.4 mm = 8.4 mm.]

A total force of 220 N was then applied to the wire in equal increments of 20 N. This was achieved by adding masses to the hanger. For each increment of force, the reading on the vernier scale was recorded. From the results, the graph shown in fig. 3.5 was drawn and conclusions were made.

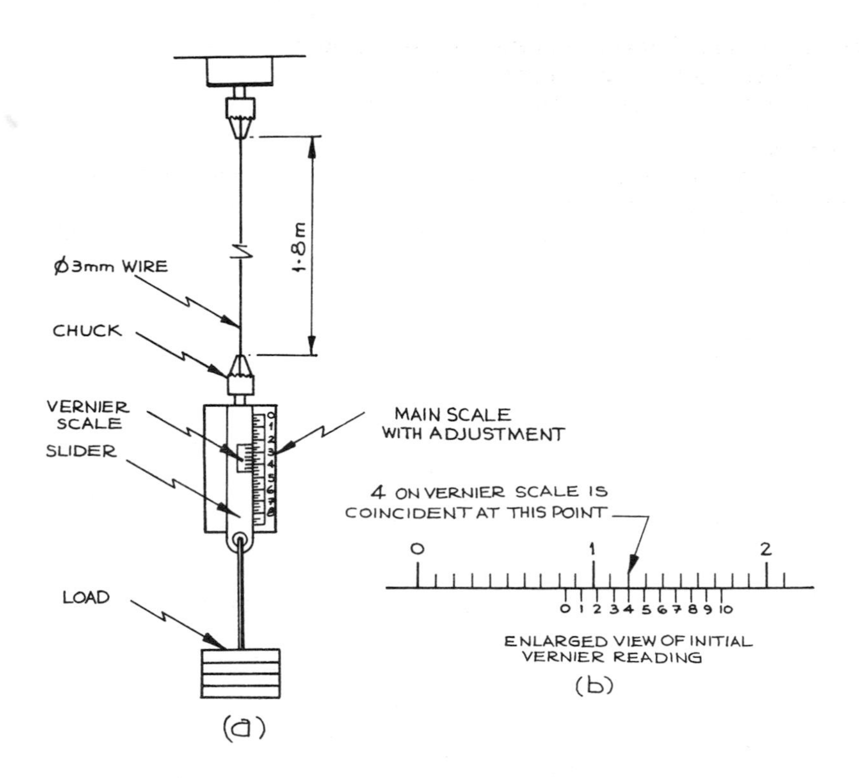

Fig. 3.4 Hooke's-law apparatus

Observations

Material Steel wire
Initial length 1.8 m
Initial diameter 3 mm

[This information is required in case it is necessary to repeat the experiment at a later date.]

Results

The results are shown in Table 3.2.

Conclusions

Referring to the graph in fig. 3.5, between O and A the extension is directly proportional to the applied force producing it. Beyond point A, the extension ceases to be proportional to the applied force – thus, point A may be termed the *limit of proportionality* [In some materials it is also the point where the material ceases to be elastic and, where this is the case, point A is then also known as the *elastic limit* for the material.]

Table 3.2 Force–extension results for a steel wire

Applied force (N)	Scale reading (mm)	Difference from first reading	=	Extension (mm)
0	8.4	8.4 – 8.4	=	0
20	8.7	8.7 – 8.4	=	0.3
40	9.0	9.0 – 8.4	=	0.6
60	9.2	9.2 – 8.4	=	0.8
80	9.5	9.5 – 8.4	=	1.1
100	9.8	9.8 – 8.4	=	1.4
120	10.1	10.1 – 8.4	=	1.7
140	10.4	10.4 – 8.4	=	2.0
160	10.6	10.6 – 8.4	=	2.2
180	10.9	10.9 – 8.4	=	2.5
200	11.2	11.2 – 8.4	=	2.8
220	11.7	11.7 – 8.4	=	3.3

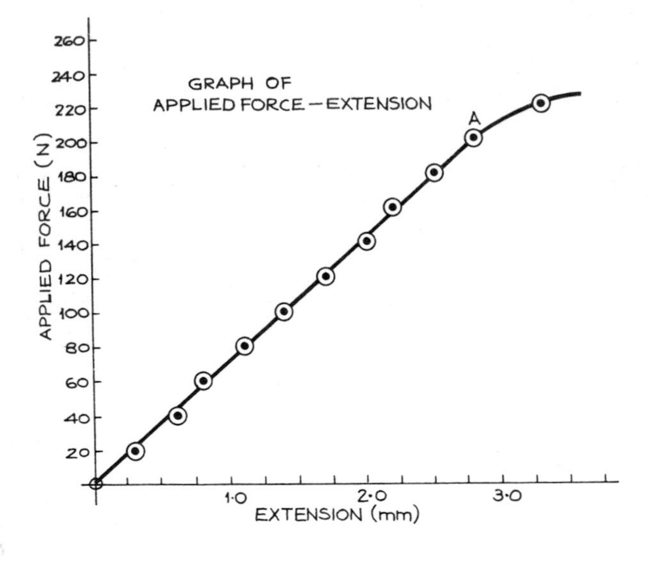

Fig. 3.5 Force–extension graph for steel wire

In carrying out experiment 3.2, care must be taken to make sure that fingers and toes are kept from beneath the suspended load. Also, it is advisable to wear safety spectacles or goggles when loading the hanger or reading the vernier scale, since any breakage or slippage of the wire (which is under tension) could cause serious injury.

3.3 Hooke's law

Experiments such as those described in section 3.2 were conducted in the seventeenth century by the English physicist *Sir Robert Hooke* (1635–1703). From his experiments, he determined that 'Within the elastic limit for a material, the extension is directly proportional to the force producing it.' *This is known as Hooke's law, and should be remembered.*

Let F = applied force

and x = extension

then, from Hooke's law,

$F \propto x$ (the symbol $\propto$ means '*proportional to*')

or $F = kx$

The constant k is known as the *stiffness* of the material and has units newtons per metre (N/m). The units newtons per millimetre (N/mm) may also be used.

If a force F_1 produces an extension x_1 and a force F_2 produces extension x_2 in the same material and Hooke's law is obeyed, then

$$k = \frac{F_1}{x_1} = \frac{F_2}{x_2} = \frac{F}{x}$$

which it is useful to remember.

It should be noted that Hooke's law may also be applied to materials in compression.

Example 1 A steel wire extends 2.2 mm when the applied force is 120 N. Assuming that the extension is within the elastic limit for the material, find the stiffness of the wire in (a) N/mm, (b) N/m.

a) $F = kx$

$$\therefore \quad k = \frac{F}{x}$$

where F = 120 N and x = 2.2 mm

$$\therefore \quad k = \frac{120\text{ N}}{2.2\text{ mm}} = 54.55\text{ N/mm}$$

i.e. the stiffness is 54.55 N/mm.

b) x = 0.0022 m

$$\therefore \quad k = \frac{120\text{ N}}{0.0022\text{ m}} = 54.55 \times 10^3\text{ N/m} \quad \text{or} \quad 54.55\text{ kN/m}$$

i.e. the stiffness is 54.55×10^3 N/m or 54.55 kN/m.

Example 2 A spring has a stiffness of 60 kN/m. Find (a) the force required to produce a compression of 3.6 mm, (b) the compression produced by a force of 360 N.

a) $F = kx$

where $k = 60\text{ kN/m} = 60 \times 10^3\text{ N/m}$ and $x = 3.6\text{ mm} = 0.0036\text{ m}$

$$\therefore \; F = 60 \times 10^3\text{ N/m} \times 0.0036\text{ m}$$

$$= 216\text{ N}$$

i.e. the force required is 216 N.

b) $F = kx$

$$\therefore \quad x = \frac{F}{k}$$

where $F = 360\text{ N}$ and $k = 60 \times 10^3\text{ N/m}$

$$\therefore \; x = \frac{360\text{ N}}{60 \times 10^3\text{ N/m}}$$

$$= 0.006\text{m or } 6\text{mm}$$

i.e. the compression induced is 6 mm.

Example 3 A copper wire extends 1.8 mm when it is subjected to a tensile force of 150 N. Assuming that the copper obeys Hooke's law, find the extension when the force is 220 N.

$$k = \frac{F_1}{x_1} = \frac{F_2}{x_2} = \frac{F}{x}$$

$$\therefore \quad x_2 = \frac{F_2}{F_1} \times x_1$$

where $F_1 = 150\text{ N}$ $F_2 = 220\text{N}$ and $x_1 = 1.8\text{ mm}$

$$\therefore \quad x_2 = \frac{220\text{N}}{150\text{N}} \times 1.8\text{mm}$$

$$= 2.64\text{mm}$$

i.e. the extension induced by a force of 220 N is 2.64 mm.

Exercises on chapter 3

1 A compression spring has a stiffness of 120 N/mm. Find the force required to compress the spring 1.6 mm.

2 A spring extends 3 mm when the applied force is 90 N. Find the extension when the force is 150 N. The spring obeys Hooke's law.

3 A wire is stretched 1.2 mm by a force of 180 N. Assuming Hooke's law applies, find the force which would stretch the wire 2 mm.
4 State Hooke's law. A compression spring has a stiffness of 30 kN/m. Find (a) the force required to compress the spring 2 mm, (b) the compression when the applied force is 110 N. The spring obeys Hooke's law.
5 Within the limit of proportionality, a tensile-test specimen extends 0.026 mm when the applied force is 12 kN. Find the extension when the applied force is 8 kN.
6 State Hooke's law. A length of copper wire is stretched 2 mm when a force of 100 N is applied. Assuming that the material obeys Hooke's law, find the extension when the applied force is (a) 50 N, (b) 150 N, (c) 160 N.
7 In an experiment on a length of steel wire, the following observations were made

Applied force (N)	50	100	150	200	250	300
Extension (mm)	0.81	1.61	2.43	3.22	4.02	4.82

Plot a graph of applied force against extension and use it to find (a) the extension when the applied force is (i) 80 N, (ii) 170 N, (iii) 260 N; (b) the applied force required to give an extension of (i) 0.4 mm, (ii) 1.2 mm, (iii) 3.4 mm.
8 The data shown below is from a tensile test on a length of steel wire. Plot the graph of force against extension, indicating on it the force at the limit of proportionality. Use the graph to find the extension produced by a force of 2.4 kN and the force required to give an extension of 0.013 mm.

Force (kN)	1.0	2.0	3.0	3.7	3.8
Extension (mm)	0.004	0.0078	0.012	0.015	0.0155

Force (kN)	3.9	4.0	4.1	4.2	4.3
Extension (mm)	0.0162	0.017	0.018	0.02	0.023

9 Figure 3.6 shows the vernier readings before and after the application of a force of 120 N. State the value of the readings shown. Determine the vernier readings when the force is increased to 150 N.

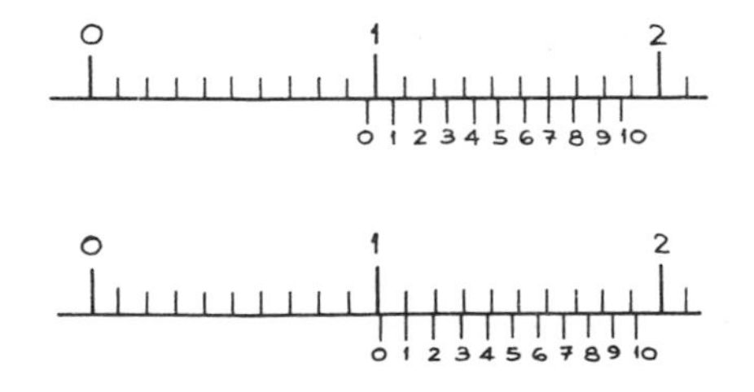

Fig. 3.6

10 In an experiment on two wires, A and B, A extends 1.1 mm when the applied force is 150 N, while B extends 0.6 mm for an applied force of 70 N.

Find the *difference* between the extensions of A and B when each wire is subjected to an applied force of 100 N.

11 Describe an experiment to determine Hooke's law. Explain clearly how the law may be determined from the data obtained.

12 Two wires, A and B, are cut from the same coil. Wire A is 0.6 m long and wire B is 1.2 m long. In a tensile test, wire A extends 0.4 mm for an applied force of 80 N. What will be the extension of wire B if the same force is applied? Both wires obey Hooke's law.

13 Four compression springs of equal length and stiffness are used to support equally a small electric motor of mass 10 kg. If the stiffness of each spring is 10 kN/m, find the compression due to the mass of the motor.

14 Two springs, each of stiffness 10 kN/m, are joined as shown in fig. 3.7. Find the total extension when the applied force at C is 50 N.

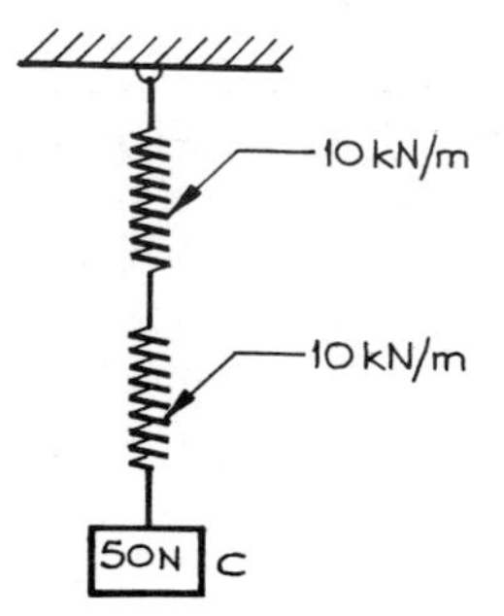

Fig. 3.7

4 Coplanar forces

4.1 Scalar and vector quantities

A *scalar* quantity has *magnitude* or *size* only. The following are scalar quantities since, in each case, the quantity has magnitude or size only:

a mass of 30kg,
a temperature of 50°C,
a distance of 2km,
a speed of 56km/h.

A *vector* quantity has both *magnitude* and *direction* and can be represented by a straight line drawn to scale. Such a line is called a *vector*.

The term *direction* is used to describe both *displacement* (which may be upwards, downwards, north or south, etc.) and *change* (which may be an increase or a decrease).

The following are vector quantities, since in each case the quantity has both magnitude *and* direction:

a temperature *increase* of 50°C,
a distance of 2km *in a north-easterly direction*,
a speed of 56km/h *due west* - this combination of speed and direction is called *velocity* (see section 6.3),
an *acceleration* of 5m/s^2 - i.e. the velocity is *increasing* at the rate of 5m/s every second.

The vector quantities above are illustrated by the vectors ***ab*** drawn to scale in fig. 4.1. Note that ***ab*** means that the direction of the vector is from a towards b.

It should be noted that *force* is also a vector quantity since, as we shall see in the next section, it is the product of a scalar quantity (mass) and a vector quantity (acceleration).

4.2 Force

Force applied to a body can cause deformation (as described in chapter 3). It can also cause or tend to cause the body to move in the direction of the force.

When a force causes a body to move, the body will accelerate, i.e. its speed will increase until friction and other resistances to motion cancel out the force's effect - the body will then continue to move with uniform speed for as long as the force is maintained.

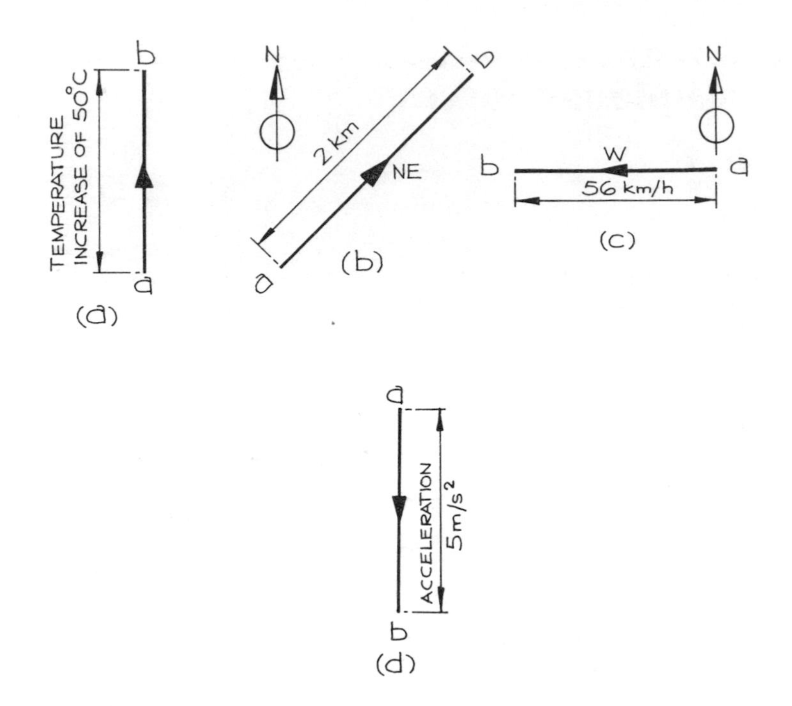

Fig. 4.1 Vectors to represent (a) a temperature increase of 50°C, (b) a distance of 2 km in a north-easterly direction, (c) a speed of 56 km/h due west, (d) an acceleration of 5 m/s^2

The magnitude of the acceleration of the body depends upon

a) the mass of the body – a small mass will accelerate more quickly than a large mass for the same applied force;
b) the applied force – a large force will produce a greater acceleration than a small force provided the mass remains the same.

From (a) above, the acceleration is *inversely proportional* to the mass. This means that, as the mass gets larger, the acceleration gets smaller provided the force remains the same. Mathematically, this may be written as

$$a \propto \frac{1}{m} \qquad \text{(i)}$$

(the symbol $\propto$ means '*proportional to*').

From (b) above, provided the mass remains unchanged, the acceleration is *directly proportional* to the applied force. This means that, as the force increases, the acceleration will also increase by a corresponding amount. Mathematically this may be written as

$$a \propto F \qquad \text{(ii)}$$

Combining equations (i) and (ii) gives

$$a \propto \frac{F}{m}$$

or $\quad F \propto ma$

$\therefore \quad F = kma$

In any system of units it is usual to let the constant $k = 1$, so that one unit of force will give one unit of acceleration to one unit of mass; thus

$$F = ma$$

which it is useful to remember.

Using SI units, the unit of mass is the kilogram (kg), the unit of acceleration is the metre per second squared (m/s^2), and the unit of force is the newton (N), which is defined as that force which will give an acceleration of one metre per second squared to a mass of one kilogram. Thus, from the above relationship, $F = ma$, we get

$$1\,N = 1\,kg \times 1\,m/s^2$$

or $\quad 1\,N = 1\,kg\,m/s^2$

which it is useful to remember.

Since force is the product of mass (a scalar quantity) and acceleration (a vector quantity), *force is a vector quantity - which should be remembered.*

When a mass is at rest or in motion on or near the earth, it is subjected to the gravitational pull of the earth (see section 6.8). The effect of the gravitational pull or force is to give or tend to give the mass a downward acceleration, denoted by the symbol g.

The gravitational force exerted on the mass is termed the *weight* (W) of the mass. The relationship between the mass (m) and the weight of a body is given by

$$W = mg$$

In Britain, the gravitational acceleration g is taken as $9.81\,m/s^2$. Hence, the downward force exerted on a mass of 1 kg is

$$W = 1\,kg \times 9.81\,m/s^2 = 9.81\,N$$

which it is useful to remember.

4.3 Moment of a force

Figure 4.2 shows a spanner being used to tighten a nut. To tighten the nut, a force must be applied to the end of the spanner in the direction shown.

When a force causes or tends to cause rotation to occur, it is applying a *turning-moment* or simply a *moment* about the centre of rotation. The moment of a force about any point is defined as the force multiplied by the perpendicular distance between the line of action of the force and the centre of rotation. Referring to fig. 4.3,

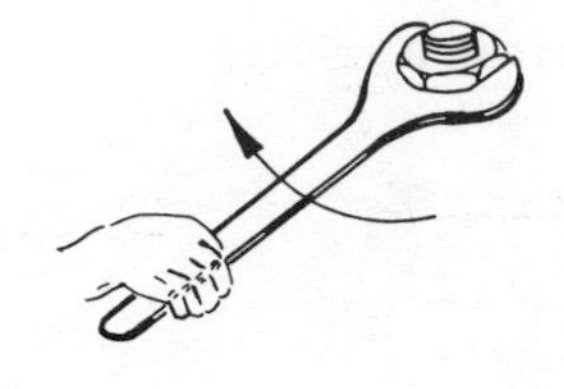

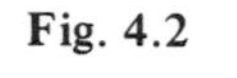

Fig. 4.2

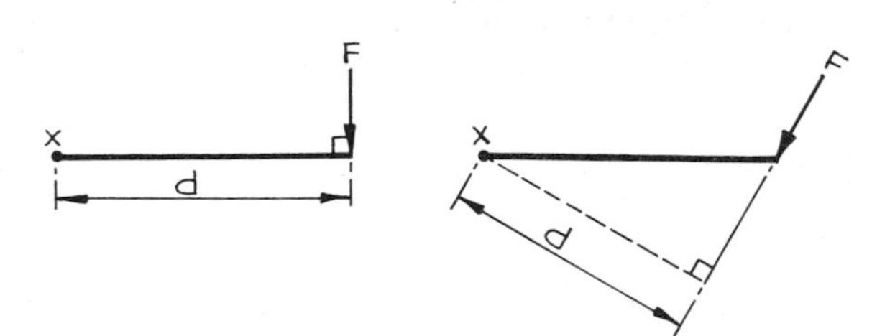

Fig. 4.3 Moment of a force

moment of force F about point X = force × perpendicular distance d between the line of action of the force and the point X

The symbol used to represent the moment of a force about any point X is M_X;

i.e. $M_X = Fd$

which should be remembered.

The units for moment of a force are newtonmetre (Nm). The units newtonmillimetre (Nmm) may also be used.

Example 1 Find the moment of the 100 N force about the point A in fig. 4.4.

$$M_A = Fd$$

where $F = 100\,\text{N}$ and $d = 200\,\text{mm} = 0.2\,\text{m}$

(In this case, d = AB since the line of action of the force is perpendicular to AB.)

$$\therefore \; M_A = 100\,\text{N} \times 0.2\,\text{m}$$

$$= 20\,\text{Nm}$$

i.e. the moment of the force about A is 20 N m.

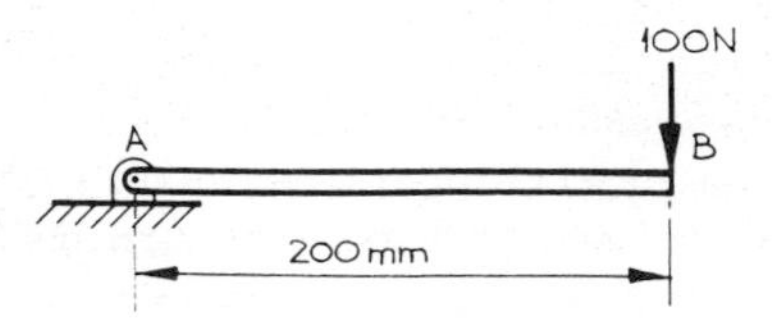

Fig. 4.4

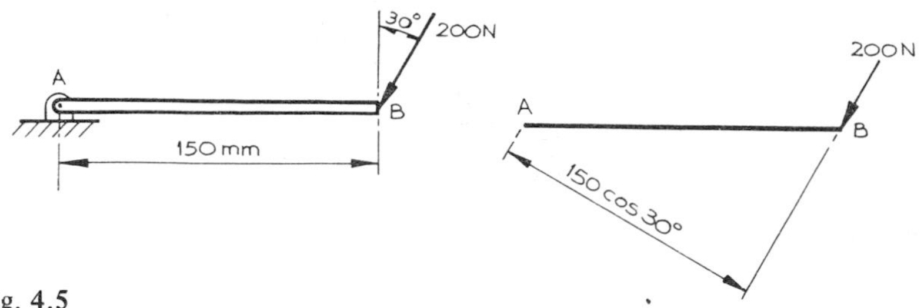

Fig. 4.5

Example 2 Find the moment of the 200 N force about the point A in fig. 4.5.

$$M_A = Fd$$

where $F = 200\text{N}$ and $d = 150\text{mm} \times \cos 30^\circ = 0.15\text{m} \times 0.866 = 0.13\text{m}$

(In this case, d – the *perpendicular distance* between the line of action of the force and A – is *not* the distance AB.)

$$\therefore \quad M_A = 200\text{N} \times 0.13\text{m}$$

$$= 26\text{Nm}$$

i.e. the moment of the force about A is 26 Nm.

Example 3 The nuts on a motor-car cylinder head require a turning-moment of 80 N m. Calculate the force which must be applied to a spanner at a distance of 450 mm from the centre of each nut.

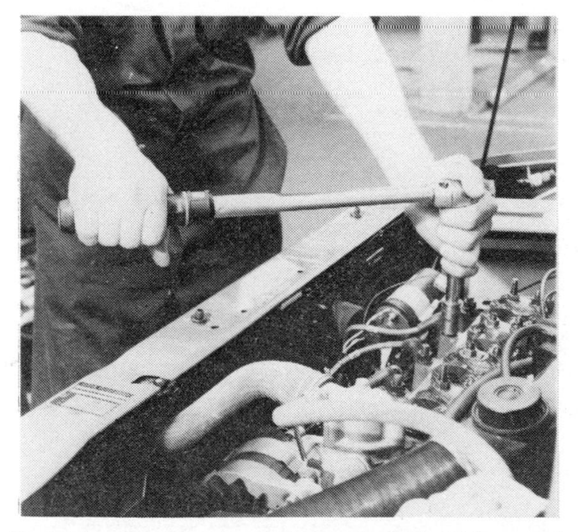

Tightening cylinder-head nuts using a torque wrench. This contains a mechanism which can be adjusted to limit the turning-effort applied to a specified value.

$$M_X = Fd$$

$$\therefore \quad F = \frac{M_X}{d}$$

where $M_X = 80\,\text{N m}$ and $d = 450\,\text{mm} = 0.45\,\text{m}$

$$\therefore \quad F = \frac{80\,\text{Nm}}{0.45\,\text{m}}$$

$$= 177.8\,\text{N}$$

i.e. the force required is 177.8 N.

4.4 Coplanar forces

When two or more forces act in the same plane, they are said to be *coplanar*.

Figures 4.6(a) and (b) show examples of coplanar forces, the forces being in the plane, or on the surface, of the paper. Figure 4.6(a) shows five parallel coplanar forces, while fig. 4.6(b) shows two coplanar forces whose lines of action pass through the *concurrent point*. The equilibrium of parallel coplanar forces only will be considered here.

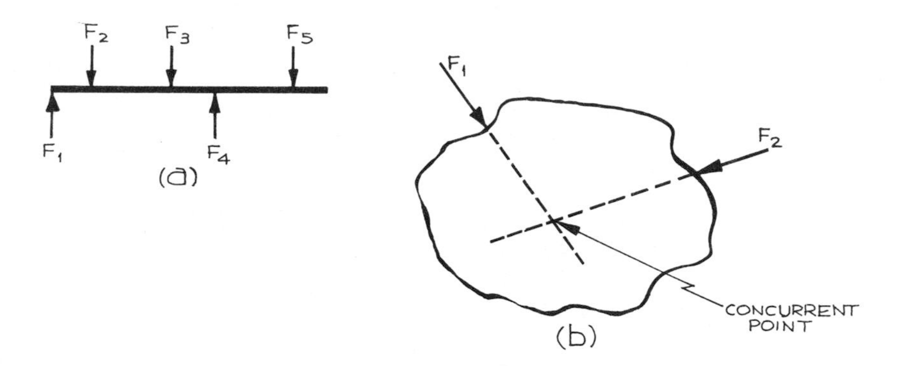

Fig. 4.6 Coplanar forces: (a) parallel coplanar forces, (b) non-parallel coplanar forces

4.4 Equilibrium of parallel coplanar forces

For any system of parallel coplanar forces to be in equilibrium – i.e. in a state of rest – two conditions must be met:

a) The sum of the forces acting in one direction must equal the sum of the forces acting in the opposite direction. For vertical and parallel coplanar forces, then,

$$\Sigma \text{ forces acting upwards} = \Sigma \text{ forces acting downwards}$$

which should be remembered. (Note: Σ (capital Greek *sigma*) means 'sum of'.)

b) The sum of the clockwise moments about any point must equal the sum of the anticlockwise moments about the *same* point,

i.e. $\sum$ clockwise moments about a point $=$ $\sum$ anticlockwise moments about the same point

This is known as the principle of moments and should be remembered.

Example 1 The uniform bar ABCD shown in fig. 4.7 is pivoted at B and supports vertical coplanar forces of 200 N at A and 150 N at C. If AB = 1 m and BC = CD = 0.75 m, find the vertical coplanar force required at D to keep the bar in equilibrium in the horizontal plane. Find also the force exerted by the pivot at B.

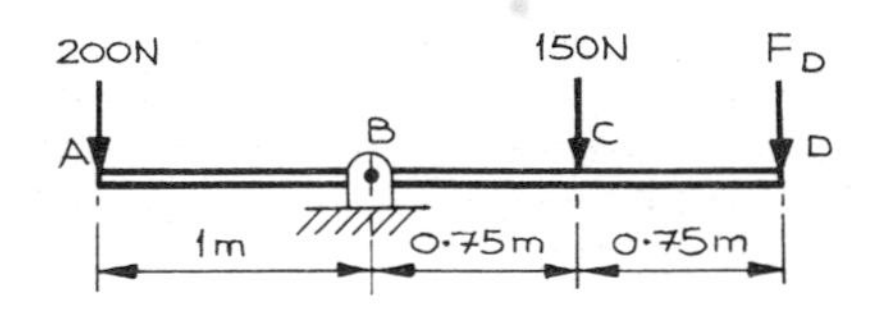

Fig. 4.7

Take moments about the pivot at B:

$\sum$ clockwise moments about a point $=$ $\sum$ anticlockwise moments about the same point

$$\therefore \quad 150\,\text{N} \times 0.75\,\text{m} + F_D \times 1.5\,\text{m} = 200\,\text{N} \times 1\,\text{m}$$

$$\therefore \quad F_D = \frac{200\,\text{Nm} - 112.5\,\text{Nm}}{1.5\,\text{m}}$$

$$= 58.3\,\text{N}$$

i.e. the force at D is 58.3 N.

$$\Sigma \text{ upward forces} = \Sigma \text{ downward forces}$$

$$\therefore \quad F_B = 200\,\text{N} + 150\,\text{N} + 58.3\,\text{N}$$

$$= 408.3\,\text{N}$$

i.e. the force exerted by the pivot at B is 408.3 N.

Example 2 The simple beam shown in fig. 4.8 is in equilibrium under the action of the parallel coplanar forces shown. Find the magnitude of the force at B.

To solve this problem, moments may be taken about A, C, or D. Take moments about A:

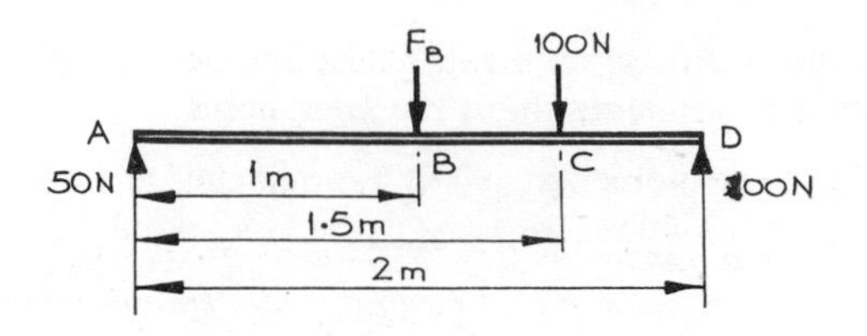

Fig. 4.8

$$\sum \begin{matrix}\text{clockwise moments}\\ \text{about a point}\end{matrix} = \sum \begin{matrix}\text{anticlockwise moments}\\ \text{about the same point}\end{matrix}$$

$\therefore \quad F_B \times 1\,\text{m} + 100\,\text{N} \times 1.5\,\text{m} = 200\,\text{N} \times 2\,\text{m}$

$\therefore \quad F_B = 250\,\text{N}$

i.e. the force at B is 250 N.
It should be noted that the 50 N force has no moment about A.

Example 3 Two metal spheres, having masses of 0.4 kg and 0.6 kg respectively, are located with their centres 300 mm apart on opposite ends of a stiff wire as shown in fig. 4.9. Neglecting the mass of the wire, find (a) the position of the pivot point so that the assembly will remain in equilibrium in the horizontal plane, (b) the total downward force on the pivot.

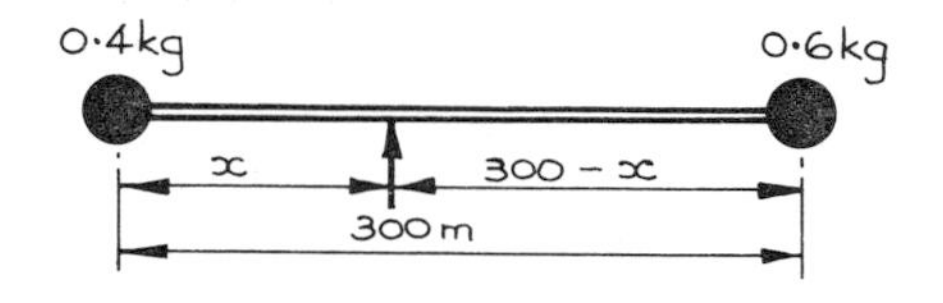

Fig. 4.9

a) Let $\quad x$ = distance from the centre of the 0.4 kg mass to the pivot

then $\quad (300\,\text{mm} - x)$ = distance from centre of the 0.6 kg mass to the pivot

Take moments of the masses of the spheres about the pivot point:

$$\sum \begin{matrix}\text{clockwise moments}\\ \text{about a point}\end{matrix} = \sum \begin{matrix}\text{anticlockwise moments}\\ \text{about the same point}\end{matrix}$$

$0.6\,\text{kg} \times (300\,\text{mm} - x) = 0.4\,\text{kg} \times x$

$\therefore \quad 180\,\text{kgmm} - 0.6\,\text{kg} \times x = 0.4\,\text{kg} \times x$

$\therefore \quad 1\,\text{kg} \times x = 180\,\text{kgmm}$

$\therefore \quad x = 180\,\text{mm}$

i.e. the pivot point is 180 mm from the centre of the 0.4 kg mass.

b) Total downward force on the pivot = downward force due to the mass of the spheres

Now a mass of 1 kg exerts a downward force of 9.81 N,

$$\therefore \quad \text{downward force on the pivot} = (0.4\,\text{kg} + 0.6\,\text{kg}) \times 9.81\,\text{N/kg}$$

$$= 9.81\,\text{N}$$

i.e. the total downward force on the pivot is 9.81 N.

In this example, since the assembly is in equilibrium about the pivot point and the total downward force is acting through this point, the pivot is located at the *centre of gravity* of the system.

4.5 Centre of gravity

All objects, large and small, on or near the surface of the earth are subjected to gravitational force (see also section 6.8). The line of action of the gravitational force acting on a body passes through a point in the body known as the centre of gravity. The *centre of gravity* of any body is defined as the point through which the line of action of the downward force due to the total mass of the body acts *in whatever position the body may be placed*. For example, the assembly shown in fig. 4.9 will remain in equilibrium in any position on the pivot while the pivot is located at the centre of gravity.

In three-dimensional bodies, the centre of gravity is also known as the *centre of mass* or *mass centre*. For plane laminae - i.e. regular or irregular shapes made from thin sheet materials - the centre of gravity is also known as the *centre of area* or the *centroid* of the area. The centre of gravity of plane laminae can be found experimentally.

Experiment 4.1 To verify that the centre of gravity of a rectangular lamina is at the intersection of the diagonals.

Take a rectangular piece of stiff card and draw on it two lines which pass through diagonally opposite corners as shown in fig. 4.10. With a sewing needle, pierce a hole at the intersection of the diagonals and thread through

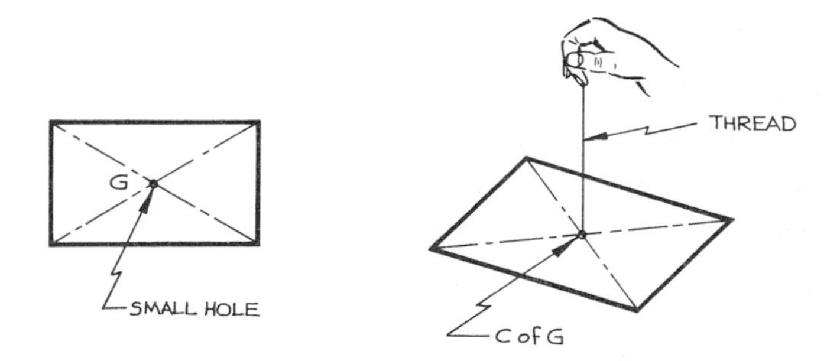

Fig. 4.10 Centre of gravity of a rectangular lamina

it a length of cotton or nylon thread, tying a knot on the free end to prevent the thread from being pulled through the hole.

Place the card on a flat surface and then lift it gently so that the card remains balanced in the horizontal plane. If balance is achieved, then this leads to the conclusion that 'With the thread passing through the intersection of the diagonals of the rectangle, the card remains balanced in the horizontal plane, thus verifying that the centre of gravity or centroid of a rectangular lamina is at this point.'

Experiment 4.2 To find the centre of gravity of *any* plane lamina.

Select an irregularly shaped lamina and drill three small holes in a triangular pattern, each hole being close to the edge as shown in fig. 4.11(a).

Select a small circular pin whose diameter is smaller than the drilled holes and clamp it to a table so that it overhangs the edge and is horizontal. Using one of the holes, suspend the lamina from the pin so that it hangs freely in the vertical plane as shown in fig. 4.11(b).

Tie a loop in one end of a length of cotton or nylon thread and on the opposite end attach a small object such as a small nut. Now pass the loop over the pin, sliding the thread as close to the vertical lamina as possible before allowing the thread to support the mass of the object, taking care to ensure that the object is not in contact with the lamina. With the object hanging freely and still, the thread will be truly vertical.

Taking great care, the position of the line of action of the thread should be marked on the lamina - this can be done by marking two points on the lamina, one near the top and one near the bottom of the thread, the points being joined by a straight line after the lamina has been removed from the pin.

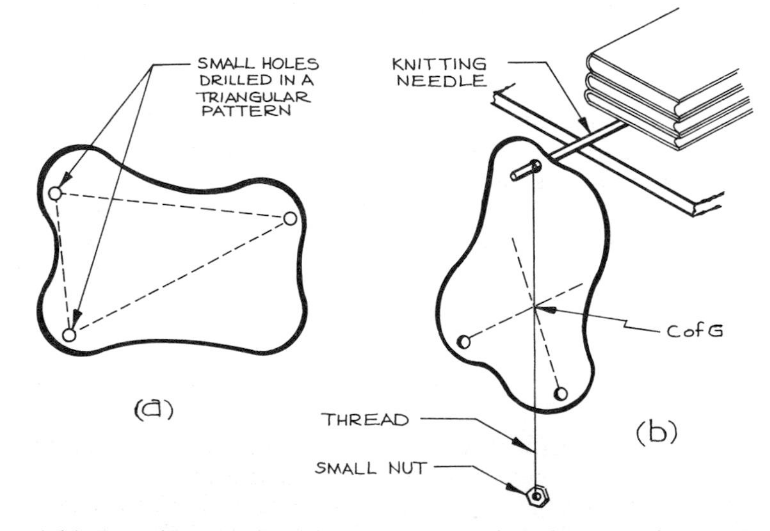

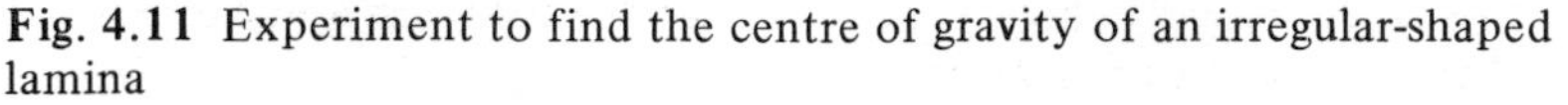
Fig. 4.11 Experiment to find the centre of gravity of an irregular-shaped lamina

This procedure should be repeated using the other two holes. If sufficient care has been taken, it will be seen that the three lines coincide or cross at a single point, this point being the centre of gravity G of the lamina.

In experiment 4.2, with the lamina suspended freely from the pin, the line of action of the total downward force due to the mass of the lamina passes through both the pin and the centre of gravity.

4.6 Types of equilibrium

The equilibrium of bodies which are at rest may be described as *stable*, *unstable*, or *neutral*. The type or nature of the equilibrium of a body is determined by the direction of the displacement of its centre of gravity G when a force is applied to the body and is then removed. This shown in Table 4.1 and illustrated in fig. 4.12.

Table 4.1 Types of equilibrium

	Displacement of the centre of gravity G	
Type of equilibrium	Force applied	Force removed
Stable	Rises	Returns to initial position
Unstable	Falls	Continues to fall
Neutral	Remains in same plane	Remains in same plane

Referring to fig. 4.12, the equilibrium of the square box and of the cone at rest on its base is *stable*, since the centre of gravity G of both bodies is *raised* when a force is applied in the direction shown and will *return to its initial position* when the force is removed. The equilibrium of the rectangular block sitting on its narrow edge and of the cone at rest on its pointed end is *unstable*, since the centre of gravity G of both bodies is *lowered* and both will *fall over* when a force is applied in the direction shown. The equilibrium of the roller or ball is *neutral*, since the centre of gravity G *remains in the same plane* whether a force is applied or not.

Also shown in fig. 4.12 is a motor-cyclist 'heeling-over' as the combination rounds a bend and a motor car being driven on its inner wheels by a stunt driver. By definition, the equilibrium of the motor-cyclist is *unstable* and that of the motor car is *stable*.

Exercises on chapter 4

1 What is meant by (a) a scalar quantity, (b) a vector quantity? Give *two* examples of each.

2 For each of the following, state whether it is a vector or a scalar quantity: time, force, speed, density, acceleration, velocity, temperature.

3 Define the term *moment of a force*. Calculate the force which must be applied to a spanner at a distance of 200 mm from the centre of a nut to give the nut a turning-moment of 30 N m.

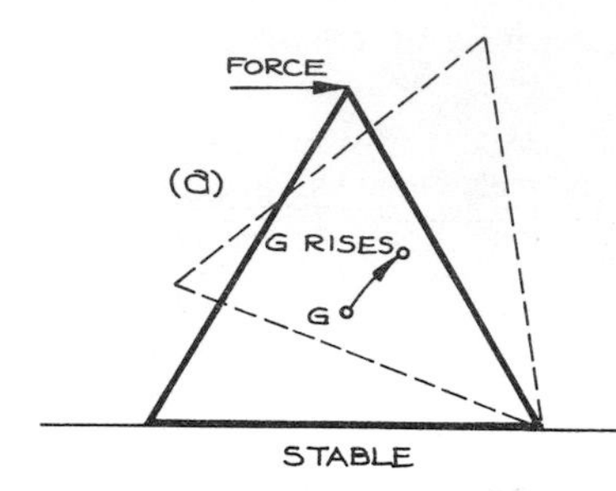

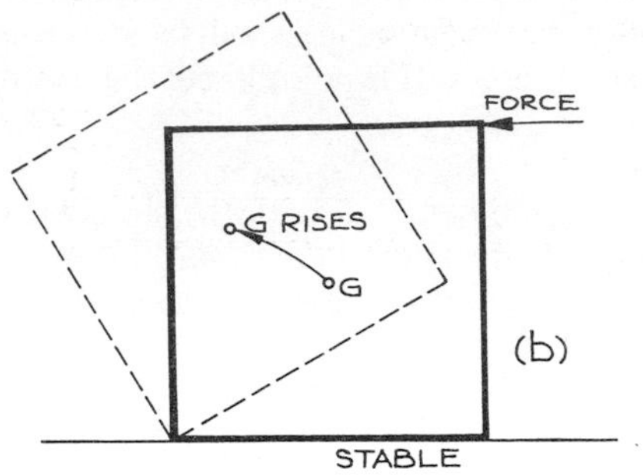

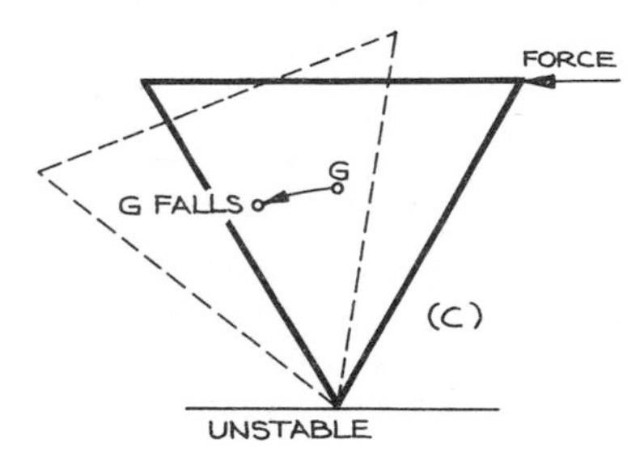

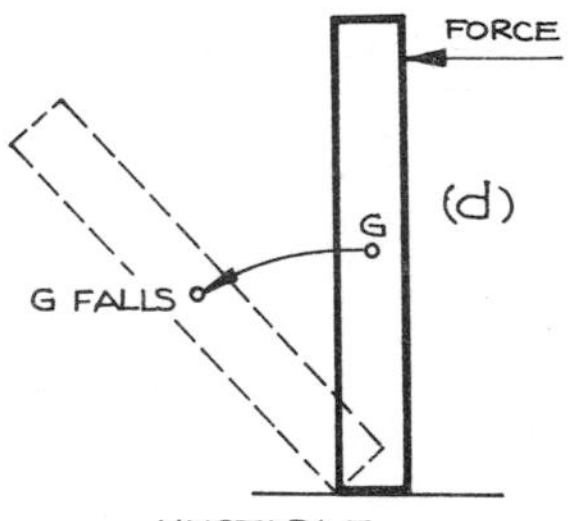

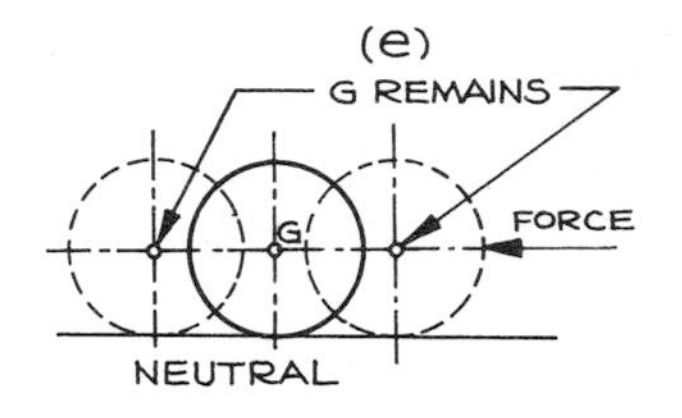

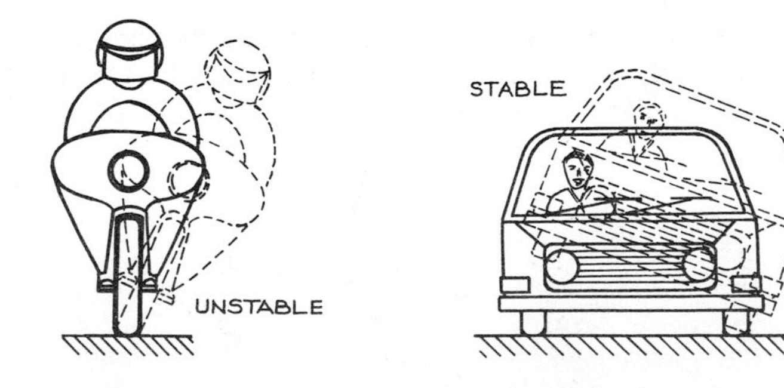

Fig. 4.12 Types of equilibrium: (a) and (b) stable, (c) and (d) unstable, (e) neutral

4 Find the moment of the force(s) about A in each of the cases shown in figs 4.13(a) to (d).

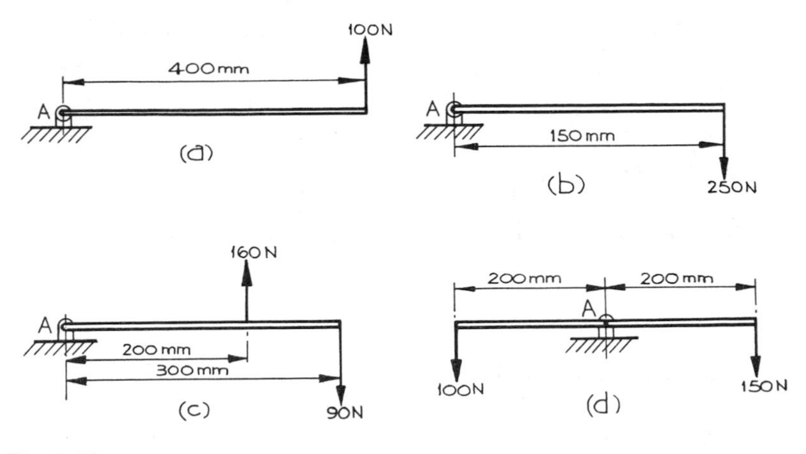

Fig. 4.13

5 Find the force required at B to keep the beam shown in fig. 4.14 in the horizontal position.

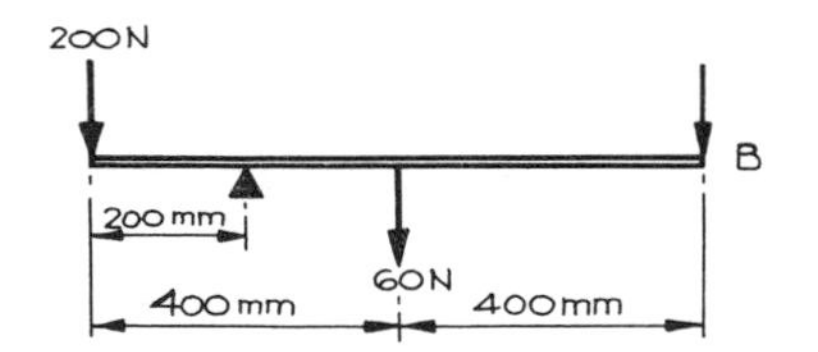

Fig. 4.14

6 The rigid bar ABC shown in fig. 4.15 is supported by springs at A and C. The stiffness of the spring at A is 30 N/mm and that of the spring at C is 60 N/mm. Calculate the extension of each spring when the force applied at B is 200 N.

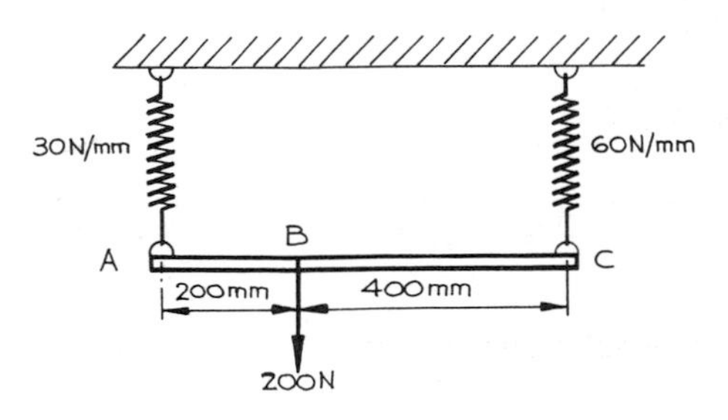

Fig. 4.15

7 Define the term *centre of gravity*. Describe an experiment to determine the centre of gravity of an irregular sheet.

8 With the aid of sketches, describe stable, unstable, and neutral equilibrium.

9 Find the centre of area of the plane figure shown in fig. 4.16. Give your answer relative to the edges AB and BC.

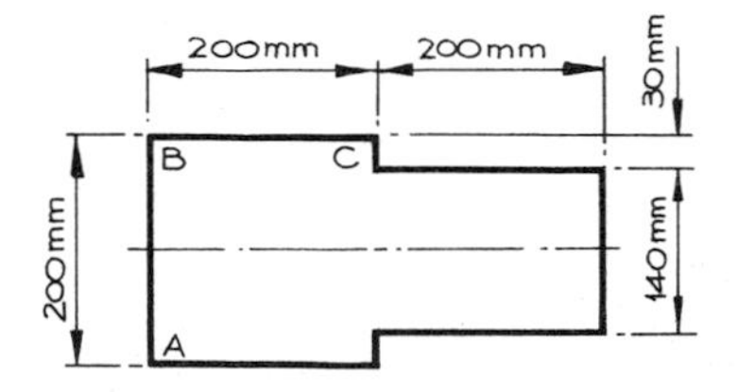

Fig. 4.16

10 A force of 60 N is applied to the handbrake shown in fig. 4.17. Find the force in the cable.

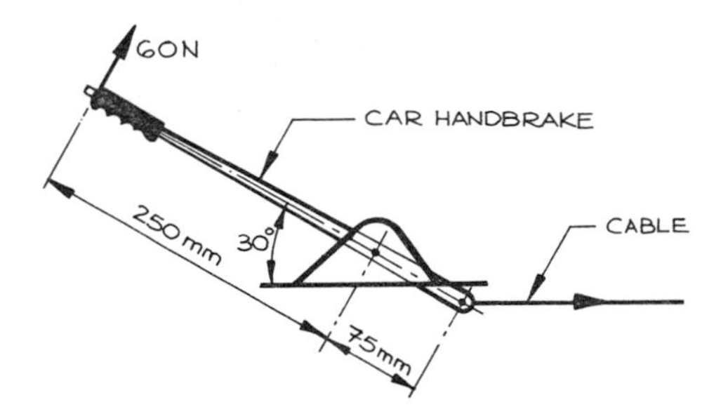

Fig. 4.17

11 Find the moment of the force(s) about A in each of the cases shown in figs 4.18 (a) to (e).

12 Find the value of the force at F required to keep the beam shown in fig. 4.19 in the horizontal plane.

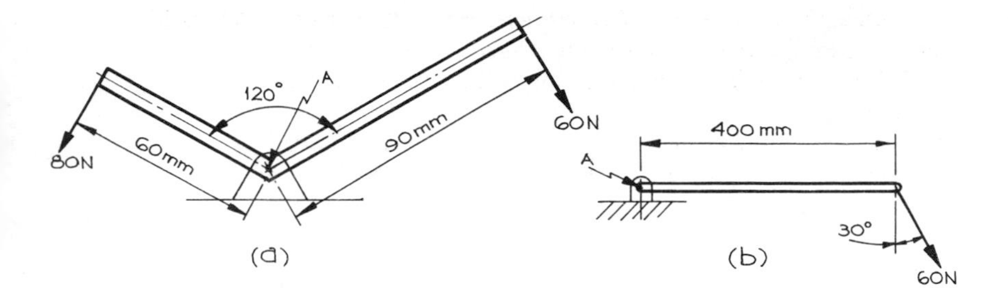

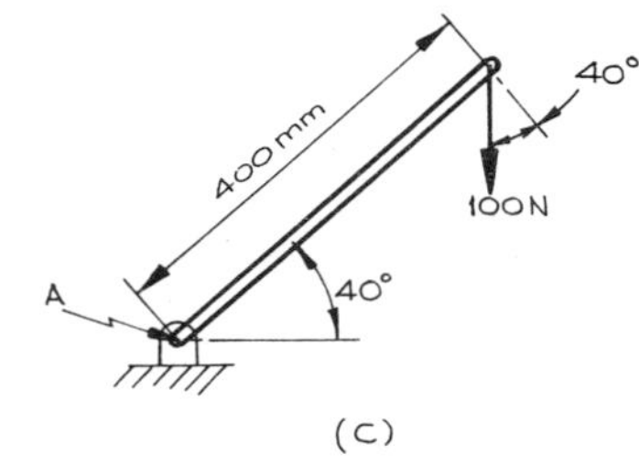

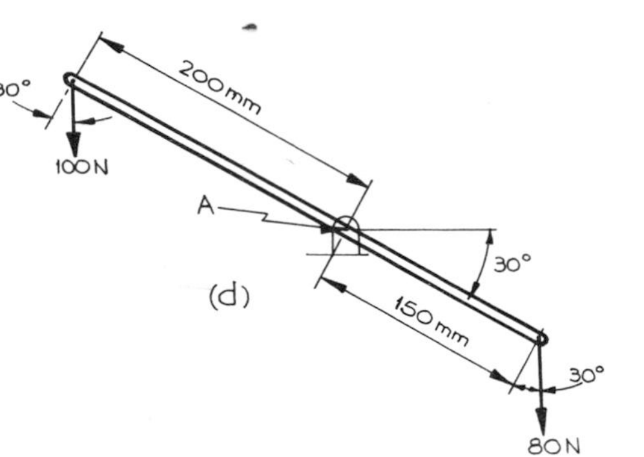

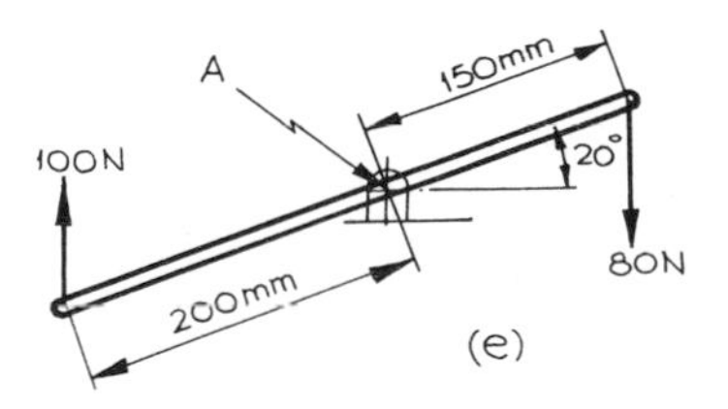

Fig. 4.18

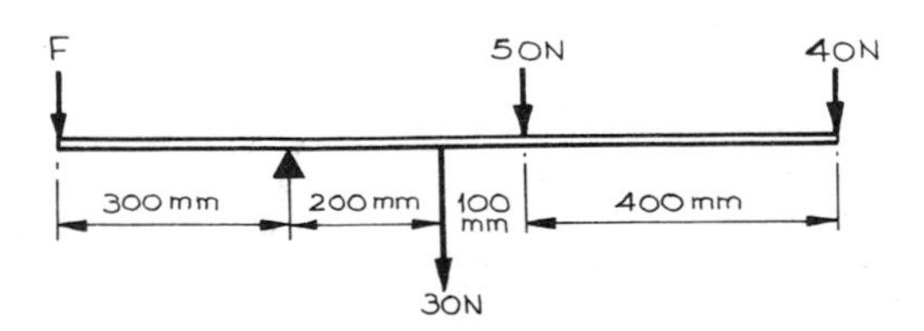

Fig. 4.19

13 By taking moments about the rear wheels, find the vertical reaction at the front wheels of the motor car shown in fig. 4.20.

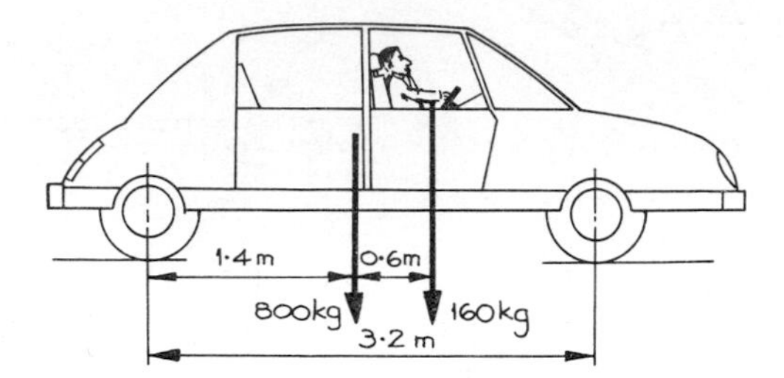

Fig. 4.20

14 State the principle of moments. The bar shown in fig. 4.21 is used to lift the crate of mass 200 kg. Calculate the force which must be applied at point A.

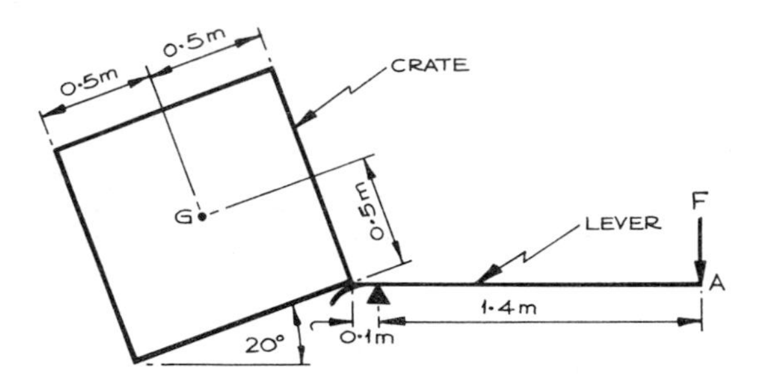

Fig. 4.21

5 Pressure in fluids

5.1 Pressure

When a motor-car tyre is inflated, air is pumped through a non-return valve into the tyre, inflation being completed when the tyre feels hard. The tyre is hard because the air inside the tyre is exerting a force both on the rubber wall of the tyre and on the steel rim of the wheel.

The force per unit area exerted by the air is called the *pressure*. Thus pressure p exerted by a fluid is defined as force F per unit area A on which the force acts

i.e. $$\text{pressure} = \frac{\text{force}}{\text{area}}$$

or $$p = \frac{F}{A}$$

which should be remembered.

The units for pressure are newtons per square metre (N/m^2). The units newtons per square millimetre (N/mm^2), pascal (Pa), and bar are also used, and it should be noted that

$$1\ \text{Pa} = 1\ N/m^2$$

and $$1\ \text{bar} = 10^5\ N/m^2 = 100000\ N/m^2 = 100\ kN/m^2$$

The 'bar' unit is particularly useful since 1 bar is approximately atmospheric pressure (standard atmospheric pressure - the pressure at which water will start to boil when its temperature is 100°C - is 1.013 25 bar). Weather charts in newspapers and on television show pressure in millibars (mbar) - standard atmospheric pressure is 1013.25 mbar.

In the water-supply industry, pipeline pressure is often measured in terms of *head* or *height* of liquid, the head being the vertical distance between the point in the pipeline at which the pressure of the water is required and the surface of the water in the storage or water tower. It is useful to note that standard atmospheric pressure (i.e. 1.01325 bar) corresponds to a height or head of 760 mm mercury (written 760 mm Hg) or 10.33 m of water (written 10.33 mH_2O);

i.e. $$1.01325\ \text{bar} \equiv 760\ \text{mm Hg} \equiv 10.33\ mH_2O$$

Example 1 The support piston in a hydraulic motor-car jack has a cross-sectional area of 500 mm^2. Calculate the pressure in the hydraulic fluid when the jack is supporting a motor car which exerts a downward force of 4.5 kN on the jack.

$$p = \frac{F}{A}$$

where $F = 4.5\,\text{kN} = 4500\,\text{N}$ and $A = 500\,\text{mm}^2 = 500 \times 10^{-6}\,\text{m}^2$

(Notice, to convert mm^2 to m^2 it is necessary to multiply by 10^{-6}.)

$$\therefore \quad p = \frac{4500\,\text{N}}{500 \times 10^{-6}\,\text{m}^2}$$

$$= 9 \times 10^6\,\text{N/m}^2 \quad \text{or} \quad 9\,\text{MN/m}^2 \quad \text{or} \quad 9\,\text{MPa} \quad \text{or} \quad 90\,\text{bar}$$

i.e. the pressure is $9 \times 10^6\,\text{N/m}^2$.

The landing gear on Concorde being supported by a hydraulic jack while the wheel is inspected

Example 2 In a gas cylinder, the cross-sectional area of the pressure-release valve is 50 mm². Find the force exerted on the valve by the gas when the pressure in the gas is 20 bar.

$$p = \frac{F}{A}$$

$$\therefore \quad F = pA$$

where $p = 20 \text{ bar} = 20 \times 10^5 \text{ N/m}^2$ and $A = 50 \text{ mm}^2 = 50 \times 10^{-6} \text{ m}^2$

$$\therefore \quad F = 20 \times 10^5 \text{ N/m}^2 \times 50 \times 10^{-6} \text{ m}^2$$

$$= 100 \text{ N}$$

i.e. the force on the valve is 100 N.

Example 3 Water is prevented from flowing from a cold-water storage tank by the valve shown in fig. 5.1. Find the minimum diameter of the valve seat required to restrict the valve closing force to 50 N when the pressure in the water is 220 kPa.

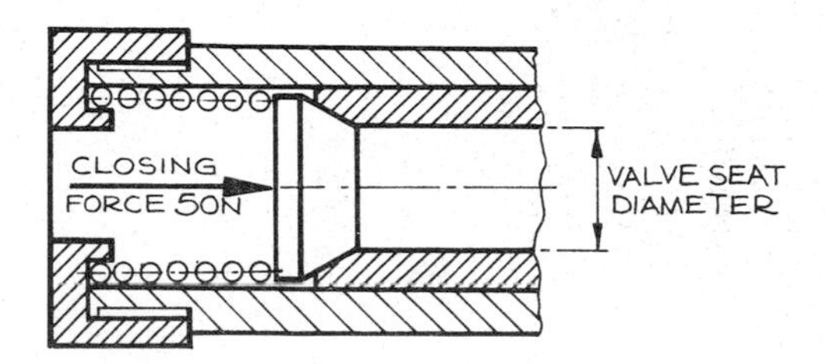

Fig. 5.1 Disc valve

$$p = \frac{F}{A}$$

$$\therefore \quad A = \frac{F}{p}$$

where $F = 50 \text{ N}$ and $p = 220 \text{ kPa} = 220 \times 10^3 \text{ N/m}^2$

$$\therefore \quad A = \frac{50 \text{ N}}{220 \times 10^3 \text{ N/m}^2}$$

$$= 227 \times 10^{-6} \text{ m}^2$$

But $A = \dfrac{\pi d^2}{4}$

where d = diameter and $\pi = 3.142$

$$\therefore \quad d = \sqrt{\frac{4A}{3.142}}$$

$$= \sqrt{\frac{4 \times 227 \times 10^{-6}\,\text{m}^2}{3.142}}$$

$$= 17 \times 10^{-3}\,\text{m} \quad \text{or} \quad 17\,\text{mm}$$

i.e. the minimum diameter of the valve seat required is 17 mm.

5.2 Absolute and gauge pressure

If the control valve on a gas cylinder is opened and the gas inside is allowed to escape to the atmosphere, the cylinder will eventually contain gas which is at the pressure of the surrounding air. If the remaining gas is now sucked out and the control valve is closed, the space inside the cylinder will be a vacuum. Where vacuum conditions exist, there is no gas present and thus zero pressure. This zero pressure is known as *absolute* zero, and pressure measured using absolute zero as a datum is called *absolute pressure*.

Generally, absolute pressure is not measured directly, since, on most pressure gauges, zero on the scale has been made to correspond to atmospheric pressure. Pressure measured from this zero point (i.e. from atmospheric pressure) is known as *gauge pressure*.

Where absolute pressure is required, it may be found from the relationship

absolute pressure = atmospheric pressure + gauge pressure

which it is useful to remember.

Example A pressure gauge attached to a gas cylinder containing acetylene indicates a pressure of 7.8 bar. Find the absolute pressure in the acetylene if the atmospheric pressure is 990 mbar.

$$\text{Absolute pressure } (p_{\text{abs.}}) = \text{atmospheric pressure } (p_{\text{at.}}) + \text{gauge pressure } (p_{\text{g}})$$

where $p_{\text{at.}} = 990\,\text{mbar} = 0.99\,\text{bar}$ and $p_{\text{g}} = 7.8\,\text{bar}$

$$\therefore \quad p_{\text{abs.}} = 0.99\,\text{bar} + 7.8\,\text{bar}$$

$$= 8.79\,\text{bar}$$

i.e. the absolute pressure in the acetylene is 8.79 bar.

5.3 Pressure in a fluid

A fluid may be a liquid or a gas. Fluids flow and take up the interior shape of their containing vessels.

Because fluids flow, they exert forces on the sides, base, and – in the case of a gas – the top of the vessel, the direction of the force always being normal to (i.e. at 90° to) the containing surface.

In a liquid, the magnitude of the force exerted at any depth below the surface of the liquid depends upon the pressure in the liquid at that point.

To show that pressure in a liquid increases with depth, observe the effect of submerging a rubber ball in a bucket of water and then releasing it. Upon release, the ball will rise quickly to float on the surface of the water. When the ball is floating, it is in equilibrium since the downward force being exerted by the ball is balanced by the upward force induced by the pressure in the water. When the ball is submerged below the surface of the water, it will be noticed that the force required to push the ball downwards increases with depth, because the pressure in the water increases; so, when the ball is released, it moves back to the surface.

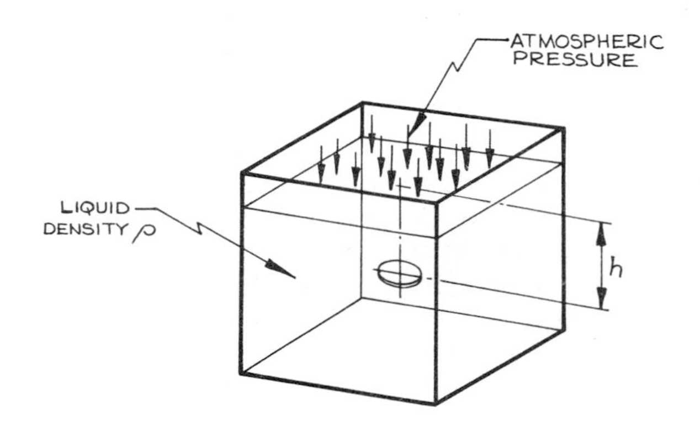

Fig. 5.2

Consider the open-topped container shown in fig. 5.2 which contains liquid of density ρ (*rho*). The absolute pressure at the surface of the liquid is the pressure of the atmosphere - i.e. the *gauge pressure* at the surface is zero. At any depth h, the gauge pressure is p_h. The magnitude of the gauge pressure p_h will depend upon

a) the depth h - the greater the depth, the larger the pressure;
b) the density ρ of the liquid - the greater the density, the larger the pressure.

Thus, the gauge pressure in the liquid is directly proportional to the depth and to the density of the fluid. This may be written mathematically in the form

$p_h \propto \rho h$ (the symbol $\propto$ means '*proportional to*')

or $p_h = \rho h g$

which it is useful to remember.

The constant g is the acceleration due to gravity (see section 6.8) and has a value of $9.81\,\text{m/s}^2 \equiv 9.81\,\text{N/kg}$.

At depth h, *the pressure p_h is the same in all directions*. However, where the liquid contacts the surface of a submerged object or the sides of the containing vessel, the direction of the pressure is *normal* (i.e. at 90°) to the surface as shown in fig. 5.3.

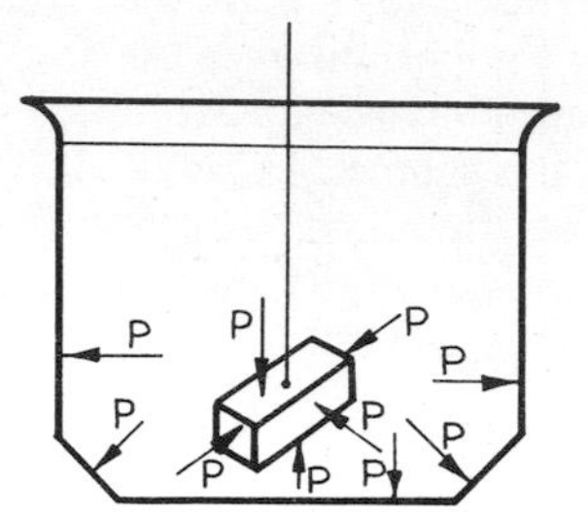

Fig. 5.3 Diagram showing that the direction of the pressure in a fluid is always normal to any surface

Example In a mercury barometer, at a temperature of 0°C, a height of 760 mm of mercury represents a pressure of 1.01325 bar. At 0°C, the density of mercury is 13 595 kg/m³. Calculate the value of the constant g.

$$p_h = \rho h g$$

$$\therefore \quad g = \frac{p_h}{\rho h}$$

where $p_h = 1.01325 \text{ bar} = 1.01325 \times 10^5 \text{ N/m}^2 \qquad \rho = 13\,595 \text{ kg/m}^3$

and $h = 760 \text{ mm} = 0.76 \text{ m}$

$$\therefore \quad g = \frac{1.01325 \times 10^5 \text{ N/m}^2}{13\,595 \text{ kg/m}^3 \times 0.76 \text{ m}}$$

$$= 9.807 \text{ N/kg}$$

i.e. the constant g is 9.81 N/kg or 9.81 m/s².

5.4 Measuring gas pressure

The pressure of a gas contained in a vessel may be measured using either a *manometer* or a *pressure gauge*.

Manometer

In its simplest form, the manometer shown in fig. 5.4 consists of a glass tube bent into the form of a 'U' and containing a quantity of liquid. One leg of the U-tube is connected to the vessel containing the gas while the other leg is open to the atmosphere.

When the gauge pressure in the gas is zero, the liquid will be at the same level in each leg. If the pressure in the gas is increased, the liquid in the

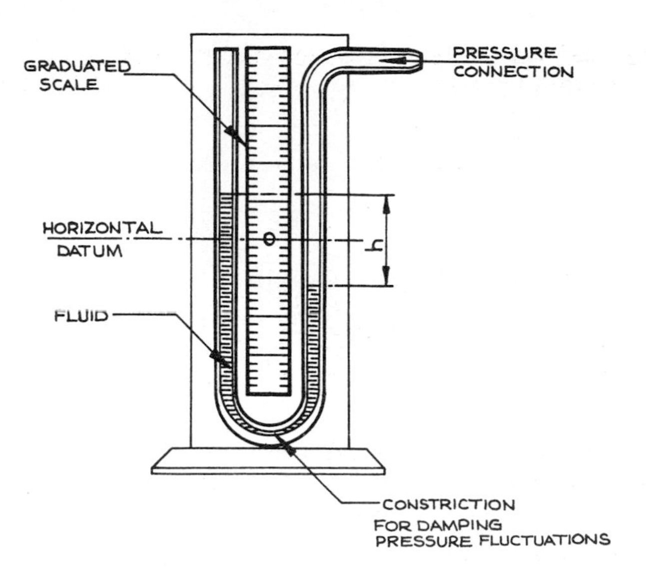

Fig. 5.4 Simple U-tube manometer

U-tube will be displaced so that there is a height difference h between the levels of the liquid in the legs. The gauge pressure p_g in the gas is given by

$$p_g = \rho g h$$

where ρ is the density of the liquid.

If the liquid used in the simple U-tube manometer is mercury, then gauge pressures up to $160 \mathrm{kN/m^2}$ may be measured. If less dense fluids are used (e.g. water), the height difference for a given pressure is greater - i.e. the *sensitivity* of the manometer is increased - but the maximum gauge pressure which can be measured with the particular manometer is reduced.

The sensitivity of a U-tube manometer can also be increased by inclining the 'open' leg as shown in fig. 5.5. The effect of inclining the 'open' leg is to increase the length of the scale. Referring to fig. 5.5, the height difference is h and the scale difference is $d = h/\sin\theta$. If, for example, $\theta = 30°$, then $\sin\theta = 0.5$ and

$$d = \frac{h}{0.5}$$

$$= 2h$$

i.e. when the 'open' leg is inclined at 30° to the horizontal, the scale difference will be twice the height difference.

Any liquid can be used in a U-tube manometer to measure the pressure of a gas, provided the liquid does not react chemically with the gas.

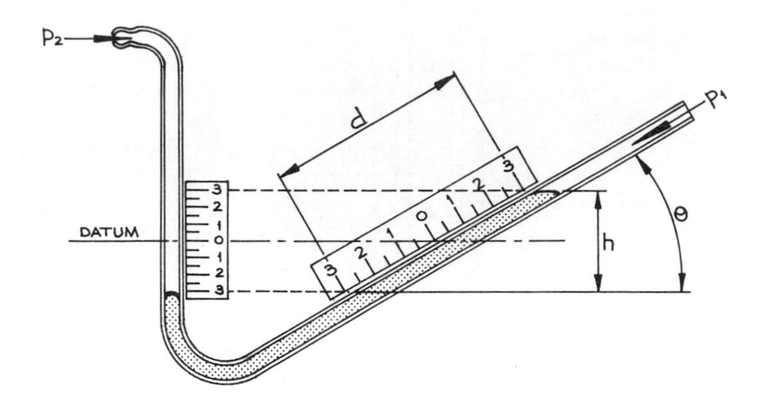

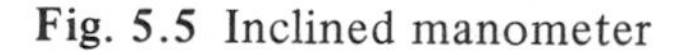
Fig. 5.5 Inclined manometer

Pressure gauge

In a pressure gauge, there is a sensing device which deforms when subjected to the pressure in the gas. The deformation, which is very small, may be converted into an electrical signal which is then amplified (i.e. made larger) or it may be magnified mechanically. In either case, the pressure causing the deformation will be indicated on a digital or analogue scale. An example of each type of scale is shown in fig. 5.6.

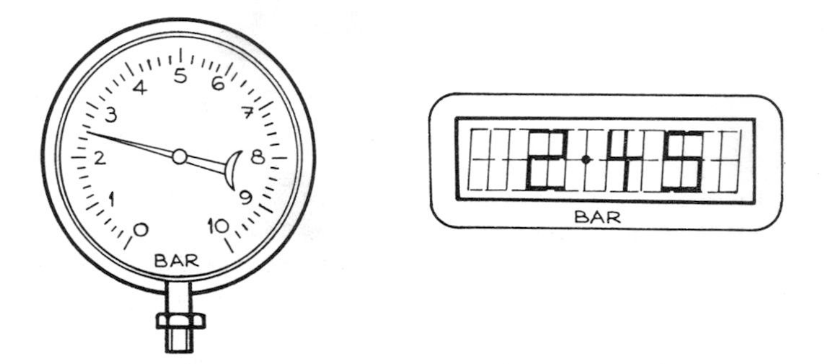

Fig. 5.6 Examples of analogue and digital scales

Pressure gauges may be obtained to measure different ranges of pressure - for example 0 to 1 bar, 0 to 5 bar, 0 to 20 bar, 0 to 50 bar etc. For greatest accuracy in measuring the pressure in a fluid within a system, the gauge with the smallest suitable range should be selected.

Exercises on chapter 5

1 Define *pressure*. The cross-sectional area of the piston in a hydraulic machine is 300 mm^2, find the force on the piston to give a fluid pressure of 12 MPa.

2 The hydraulic braking system in a motor car consists of a master-cylinder having a piston cross-sectional area of 100 mm^2 which is connected to the wheel slave-cylinders each with a piston cross-sectional area of 300 mm^2. When the applied force on the master-cylinder piston is 200 N, find (a) the pressure in the hydraulic fluid, (b) the force exerted by each slave-cylinder piston.

3 The cross-sectional area of a pressure-relief valve in a steam line is 200 mm^2. If the external force on the valve is 100 N, calculate the maximum pressure in the steam in (a) N/mm^2, (b) Pa, (c) bar.

4 In an internal-combustion engine, the cross-sectional area of each piston is 4560 mm^2. During the compression stage, the pressure in the petrol–air mixture rises to 2 MPa. Find the force exerted by each piston during compression.

5 The flask shown in fig. 5.7 contains a liquid. Copy the figure and show on it the direction of the pressure at points A, B, C, and D.

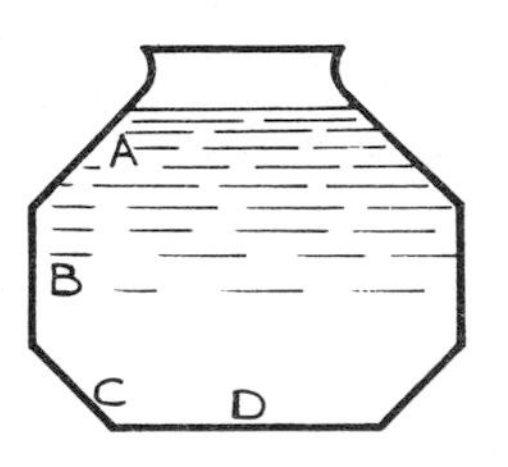

Fig.5.7

6 State the relationship between *absolute* and *gauge* pressure. The pressure in a pneumatic tyre is measured and found to be 2 bar If the atmospheric pressure is 1.01 bar, calculate the absolute pressure in the tyre in (a) bar, (b) pascal.

7 A submarine is sailing at a depth of 300 m in sea water having a density of 1020 kg/m^3. Calculate the absolute pressure on the hull of the submarine when the atmospheric pressure is 99 kPa.

8 A mercury barometer registers a pressure of 760 mm. Given that the density of mercury is 13 600 kg/m^3, convert this pressure to (a) bar, (b) pascal.

9 Find the height of the liquid in a mercury barometer when the atmospheric pressure is (a) 975 mbar, (b) 100.1 kPa, (c) 0.0992 N/mm^2. The density of the mercury is 13 600 kg/m^3.

10 A U-tube manometer contains mercury of density 13 600 kg/m^3. Calculate the measured pressure when the height difference between the fluid in the open and closed legs of the manometer is 50 mm.

11 The surface of a reservoir is 150 m vertically above the outlet nozzle to a water turbine. Calculate the pressure in the water at the outlet nozzle. The density of the water is 1000 kg/m^3.

12 In a gravity-feed house central-heating system, the surface of the water in the header tank is 6.2 m above ground level. The inlets to the ground-floor and first-floor radiators are respectively 150 mm and 2.9 m above ground level. Calculate the absolute pressure at entry to these radiators when the atmospheric pressure is 98.2 kPa. Take the density of water as 1000 kg/m^2.

13 Select a suitable pressure gauge to measure each of the following:

a) domestic natural-gas supply in the range 0 to 2 kPa,
b) compressed air in the range 0 to 7 bar,
c) steam in the range 0 to 10 MPa,
d) atmospheric pressure,
e) liquid in the range 0 to 40 kPa.

14 The inclined-tube manometer shown in fig. 5.8 contains a fluid of density 2600 kg/m^3. Find (a) the gauge pressure in the system, (b) the absolute pressure when the atmospheric pressure is 1.005 bar.

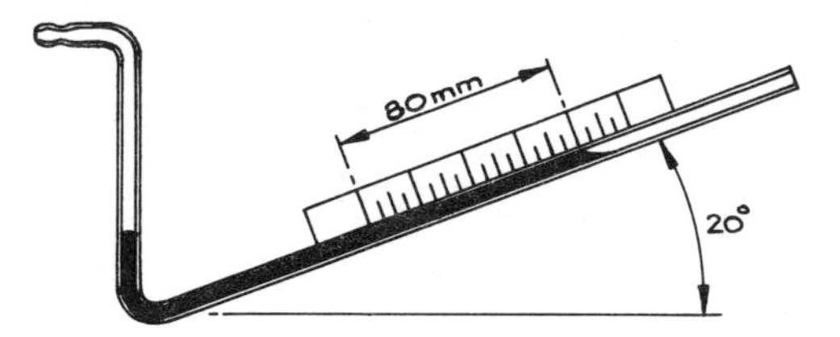

Fig. 5.8

6 Velocity and acceleration

6.1 Displacement

Figure 6.1 shows the path followed by a vehicle in moving from A to B. The distance travelled by the vehicle is 35 km, while the *displacement* of the vehicle - i.e. the straight-line distance from the start to the finish - is 20 km in a north-westerly direction.

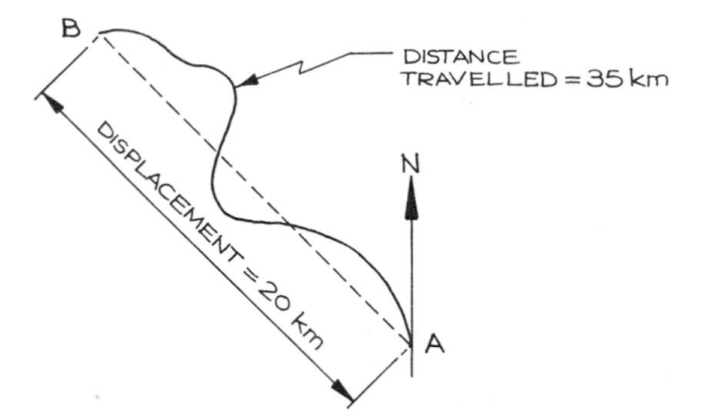

Fig. 6.1

The distance travelled is a scalar quantity, since it has magnitude only, but *displacement* is a vector quantity since it has both magnitude *and* direction.

The unit for displacement is the metre (m). The units kilometre (km) and millimetre (mm) are also used. The symbol used for displacement is s,

i.e. displacement s = distance travelled in a specified direction

which should be remembered.

Example A toy train moves around a circular track of radius 1 m. If the train moves from point A to point B which is diametrically opposite as shown in fig. 6.2, find (a) the distance travelled by the train, (b) the displacement of the train.

a) Distance travelled = length of arc between A and B

and length of arc = $\pi \times$ radius

where $\pi = 3.14$ and radius = 1 m

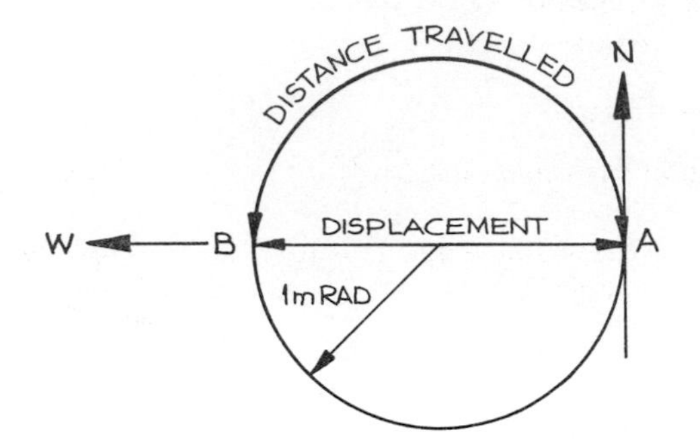

Fig. 6.2

$\therefore$ distance travelled = $3.14 \times 1\,\text{m}$

$= 3.14\,\text{m}$

i.e. the distance travelled is 3.14m.

b) Displacement = distance travelled in a specified direction

In moving from A to B, the train is displaced 2m due west from A; i.e. the displacement is 2m due west.

6.2 Speed

Speed is defined as distance travelled, s, per unit time t,

$$\text{i.e.}\quad \text{speed} = \frac{\text{distance travelled}}{\text{time taken}}$$

$$= \frac{s}{t}$$

The unit for speed is the metre per second (m/s). The units kilometre per hour (km/h), metres per minute (m/min), and millimetre per second (mm/s) may also be used.

Because speed has magnitude or size only, it is a *scalar* quantity (see section 4.1).

Example 1 A runner completes a marathon race of 42.2 km in 2 hours 36 minutes. Find the average speed of the runner.

$$\text{Average speed} = \frac{\text{distance travelled}}{\text{time taken}} = \frac{s}{t}$$

where $s = 42.2\,\text{km} = 42\,200\,\text{m}$

and $t = 2\,\text{h}\ 36\,\text{min} = 2\,\text{h} \times 3600\,\text{s/h} + 36\,\text{min} \times 60\,\text{s/min} = 9360\,\text{s}$

$$\therefore \quad \text{average speed} = \frac{42\,200\,\text{m}}{9360\,\text{s}}$$

$$= 4.5\,\text{m/s}$$

i.e. the average speed of the runner is 4.5 m/s.

Example 2 A motor car travels at a constant speed of 50 km/h for 2 minutes. Calculate the distance travelled.

$$\text{Speed} = \frac{s}{t}$$

$$\therefore \quad s = t \times \text{speed}$$

where $t = 2\,\text{min} = 120\,\text{s}$

and $\text{speed} = 50\,\text{km/h} = \dfrac{50\,\text{km/h} \times 1000\,\text{m/km}}{3600\,\text{s/h}} = 13.9\,\text{m/s}$

$$\therefore \quad s = 120\,\text{s} \times 13.9\,\text{m/s}$$

$$= 1668\,\text{m} \quad \text{or} \quad 1.668\,\text{km}$$

i.e. the distance travelled is 1.668 km.

Example 3 During a journey of 80 km completed in 1.5 h, a motor car reached a speed of 140 km/h. Calculate the average speed for the journey, in km/h, and explain the difference between this and the maximum speed.

$$\text{Average speed} = \frac{\text{distance travelled}}{\text{time taken}} = \frac{s}{t}$$

where $s = 80\,\text{km}$ and $t = 1.5\,\text{h}$

$$\therefore \quad \text{average speed} = \frac{80\,\text{km}}{1.5\,\text{h}}$$

$$= 53.3\,\text{km/h}$$

i.e. the average speed for the hourney was 53.5 km/h.

During the journey, the speed of the car varies between zero at the start and finish or at traffic hold-ups to 140 km/h. The speed at any *instant* during the journey is known as the *instantaneous speed*, while the speed for the whole journey is called the *average speed*.

6.3 Speed and velocity

Velocity, v, is defined as displacement (i.e. distance travelled in a specified direction) s per unit time t;

$$\text{i.e.} \quad \text{velocity} = \frac{\text{displacement}}{\text{time taken}}$$

or $$v = \frac{s}{t}$$

which should be remembered.

Velocity has the same units as speed. Since velocity has both magnitude (i.e. speed) *and* direction, it may be defined as *speed in a specified direction*. Velocity is thus a vector quantity and may be represented by a *vector* drawn to scale.

Figure 6.3 shows a vector representing the velocity of an aircraft flying due east with a speed of 1000 km/h.

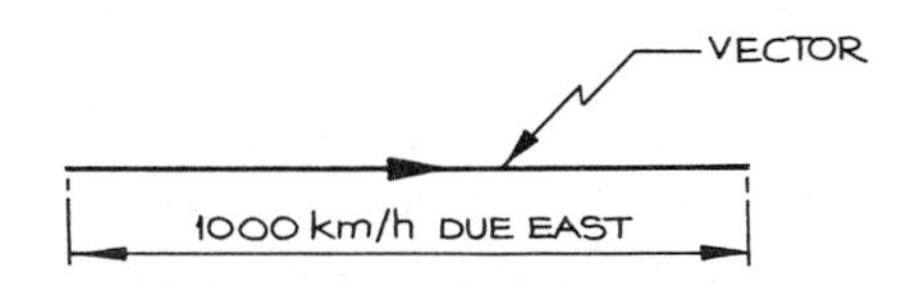

Fig. 6.3 Vector representing the velocity of an aircraft

Example 1 A ship sails north-east for a distance of 60 km in 2 hours. Calculate the average velocity in km/h.

$$\text{Average velocity} = \frac{\text{distance travelled in a specified direction}}{\text{time taken}} = \frac{s}{t}$$

where $s = 60\,\text{km}$ (in a north-easterly direction) and $t = 2\,\text{h}$

$$\therefore \quad v = \frac{60\,\text{km}}{2\,\text{h}} = 30\,\text{km/h}$$

i.e. the average velocity is 30 km/h in a north-easterly direction.

Example 2 A runner completes a marathon race of 42.4 km in 2 hours 12 minutes. If the finish of the race is 12 km due west from the start, calculate the average speed and the average velocity of the runner.

$$\text{Average speed} = \frac{\text{total distance travelled}}{\text{time taken}} = \frac{s}{t}$$

where $s = 42.2\,\text{km} = 42\,200\,\text{m}$

and $t = 2\,\text{h}\ 12\,\text{min} = 2\,\text{h} \times 3600\,\text{s/h} + 12\,\text{min} \times 60\,\text{s/min} = 7920\,\text{s}$

$$\therefore \quad \text{average speed} = \frac{42\,200\,\text{m}}{7920\,\text{s}}$$

$$= 5.33\,\text{m/s}$$

i.e. the average *speed* of the runner is 5.33 m/s.

$$\text{Average velocity} = \frac{\text{distance travelled in a specified direction}}{\text{time taken}} = \frac{s}{t}$$

In this case, $s = 12\,\text{km}$ due west

i.e. $s = 12\,000\,\text{m}$ and $t = 7920\,\text{s}$

$$\therefore \quad v = \frac{12000\,\text{m}}{7920\,\text{s}}$$

$$= 1.52\,\text{m/s}$$

i.e. the average *velocity* of the runner is 1.52 m/s due west.

Example 3 The speed of a car changes from 30 km/h to 45 km/h. Calculate the difference in speed and explain why this is a change in velocity.

$$\begin{aligned}\text{Difference in speed} &= \text{final speed} - \text{initial speed}\\ &= 45\,\text{km/h} - 30\,\text{km/h}\\ &= 15\,\text{km/h (increase)}\end{aligned}$$

This is a change in velocity since it involves both magnitude (i.e. 15 km/h) and direction (i.e. it is an increase) - i.e. *if the speed of a moving body alters, then the difference in speed is a change in velocity.*

Example 4 A model steam engine moves with a constant speed of 1.5 m/s around a circular track. Calculate the change in the velocity of the engine between two points A and B which are diametrically opposite.

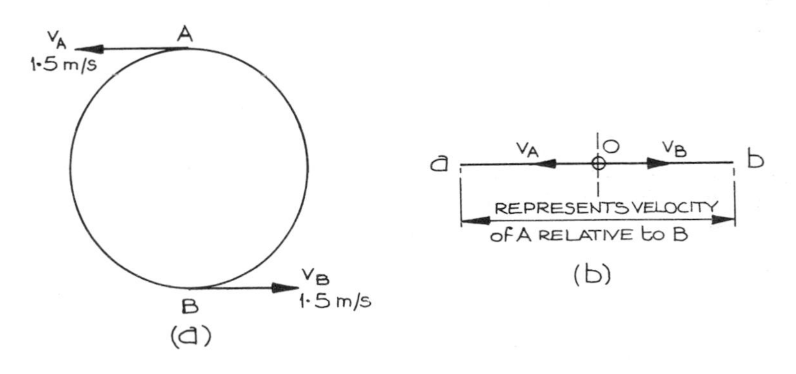

Fig. 6.4

Referring to fig. 6.4(a), at A the velocity of the engine is 1.5 m/s. This is because the engine has a speed of 1.5 m/s in a direction which is tangential to the circular track at A. This velocity is represented by the vector ***oa***, 1.5 units long, in fig. 6.4(b). At B, the velocity is 1.5 m/s and is represented by the vector ***ob***, 1.5 units long, in fig. 6.4(b).

Referring to fig. 6.4(b), notice that the directions of the vectors - i.e. o *towards* a and o *towards* b - are opposite; thus, relative to A, the velocity at B is represented by the vector ***ab*** which is 3 units long. The change in velocity between A and B is therefore 3 m/s.
i.e., *when a body moving at constant speed changes direction, then there will be a change in velocity*.

6.4 Distance-time graphs

Distance-time graphs may be plotted from data obtained in a number of ways. For example, alongside motorways there are marker posts 100 m apart which indicate the direction to the nearest emergency telephone. Using the marker posts and with the aid of a stop-watch, the results in Table 6.1 have been obtained, from which the distance-time graph shown in fig. 6.5 has been drawn.

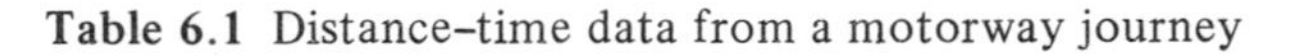

Table 6.1 Distance-time data from a motorway journey

Distance (m)	0	200	400	600	800	1000	1200	1400	1600	1800	2000
Time (s)	0	8.6	17.8	27.2	35.8	45.0	54.2	63.0	71.6	80.0	90.1

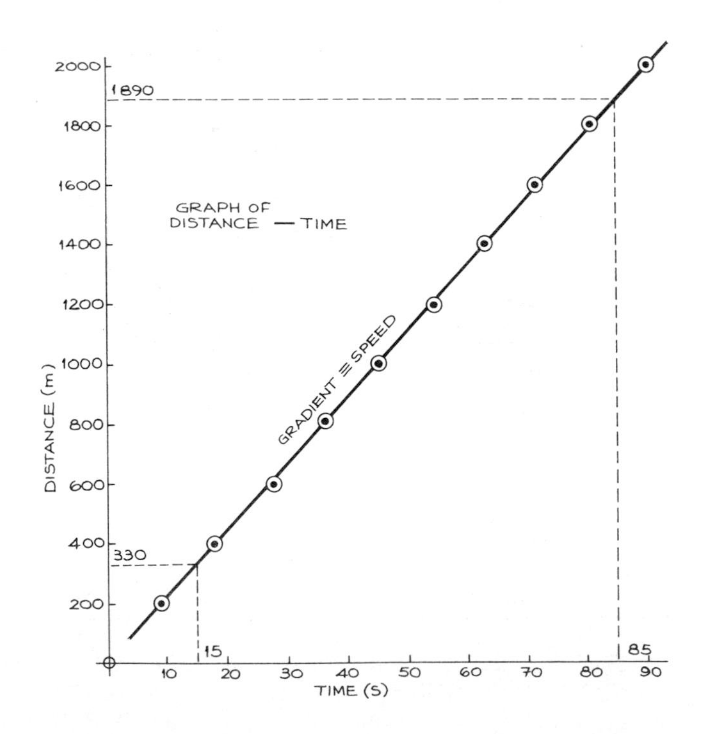

Fig. 6.5 Distance-time graph for a motorway journey

Referring to the straight-line graph shown in fig. 6.5, the *speed* of the vehicle is represented by the slope or gradient of the graph;

i.e. speed $\equiv$ gradient of the distance–time graph $= \dfrac{s}{t}$

To find the gradient, select two *non-plotted points* on the line which are as far apart as possible (this is to give greater accuracy). For each point, note the values of distance s and time t.

In this case, at $s_1 = 330\,\text{m}$, $t_1 = 15\,\text{s}$; and at $s_2 = 1890\,\text{m}$, $t_2 = 85\,\text{s}$;

$$\therefore \quad \text{gradient} = \frac{s_2 - s_1}{t_2 - t_1}$$

$$= \frac{1890\,\text{m} - 330\,\text{m}}{85\,\text{s} - 15\,\text{s}}$$

$$= 22.3\,\text{m/s}$$

i.e. the speed of the vehicle is represented by the gradient of the distance–time graph and is 22.3 m/s.

In the physical-science laboratory, distance–time data can be obtained from moving trolleys using a vibrating-pen recorder or a ticker-tape timer.

Figure 6.6 shows a vibrating-pen recorder making a trace on a moving trolley. The recorder consists of a steel strip which is rigidly held at one end and carries a small mass and pen at the other.

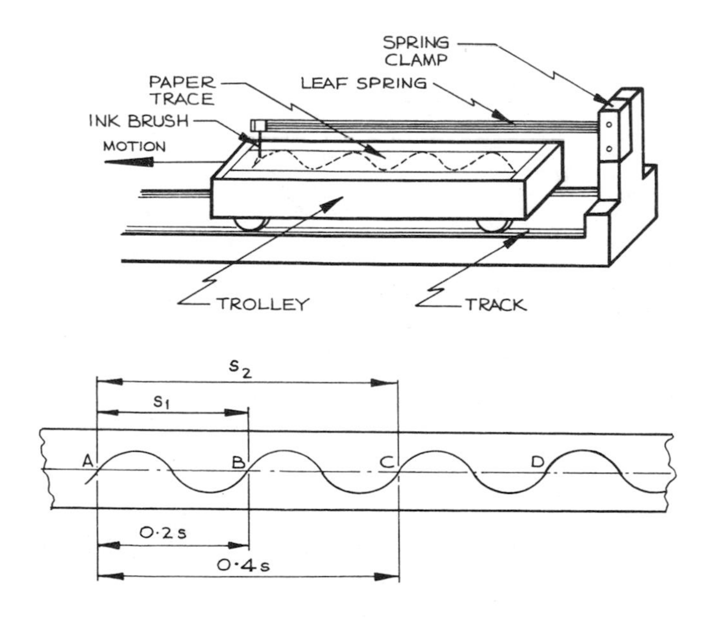

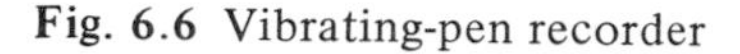
Fig. 6.6 Vibrating-pen recorder

To operate the recorder, the mass is displaced either to the left or to the right of the equilibrium position and is then released. Upon release, the mass will oscillate from side to side, the time taken for each successive oscillation being the same. If the trolley now moves under the oscillating pen, a trace similar to the one shown will be produced.

Referring to the trace, the curves between A and B, between B and C, between C and D, etc. represent one complete oscillation of the pen, while the linear distances between A, B, C, D, etc. represent the distance moved by the trolley. Thus, if the pen oscillates at say 5 times per second, then s_1 represents the distance moved by the trolley in 0.2 s, s_2 is the distance moved in 0.4 s, etc.

The results in the following example were found using a trolley and a vibrating-pen recorder.

Example An experiment on a moving trolley yielded the following values for distance and time:

Time t (s)	0	1	2	3	4	5
Distance s (mm)	0	50	95	145	195	240

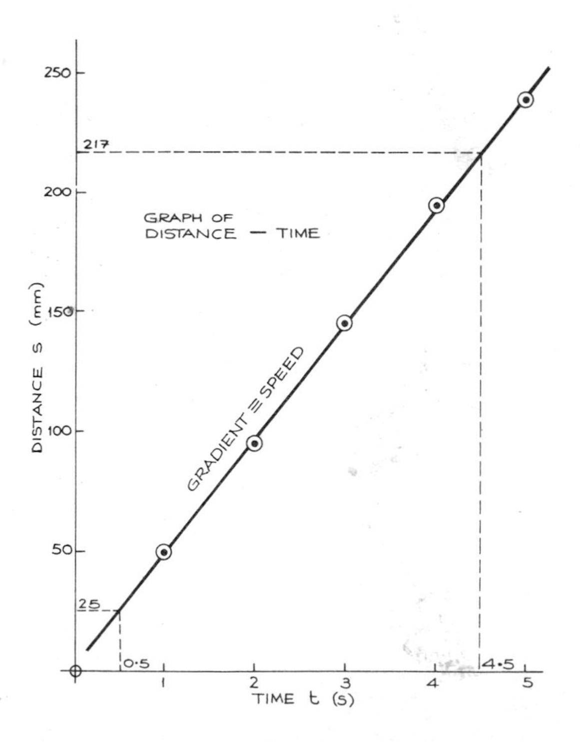

Fig. 6.7 Distance–time graph for an experiment on a moving trolley

Plot a graph of distance against time and use it to determine the speed of the trolley.

$$\text{Speed} \equiv \text{gradient of distance-time graph} = \frac{s_2 - s_1}{t_2 - t_1}$$

From the graph shown in fig. 6.7, at $s_1 = 25$ mm, $t_1 = 0.5$ s; and at $s_2 = 217$ mm, $t_2 = 4.5$ s;

$$\therefore \quad \text{gradient} = \frac{217\,\text{mm} - 25\,\text{mm}}{4.5\,\text{s} - 0.5\,\text{s}}$$

$$= 48\,\text{mm/s}$$

i.e. the speed of the trolley is represented by the gradient of the distance-time graph and is 48 mm/s.

6.5 Acceleration

Acceleration, a, is defined as change in velocity, Δv, per unit time time t;

$$\text{i.e. acceleration} = \frac{\text{change in velocity}}{\text{time}}$$

or

$$a = \frac{\Delta v}{t}$$

which should be remembered.

The unit for acceleration is the metre per second per second, usually termed metre per second squared (m/s^2). The unit millimetre per second squared (mm/s^2) may also be used.

Because acceleration depends on a change in velocity, it is a *vector* quantity. Where acceleration causes a *decrease* in velocity, it is termed *deceleration* or *retardation*

Example 1 The speed of a motor car increases from 15 km/h to 75 km/h in 6 seconds. Calculate the acceleration.

$$a = \frac{\Delta v}{t}$$

where $\Delta v = 75\,\text{km/h} - 15\,\text{km/h} = 60\,\text{km/h} = \dfrac{60\,\text{km/h} \times 1000\,\text{m/km}}{3600\,\text{s/h}}$

$$= 16.7\,\text{m/s}$$

and $t = 6\,\text{s}$

$$\therefore \quad a = \frac{16.7\,\text{m/s}}{6\,\text{s}}$$

$$= 2.78\,\text{m/s}^2$$

i.e. the acceleration is $2.78\,m/s^2$.

Example 2 The brakes of an electric locomotive are applied when it is travelling at 170 km/h, retarding it at the rate of 1.8 m/s². Calculate the time required to reduce the speed of the locomotive to 100 km/h.

$$a = \frac{\Delta v}{t}$$

$$\therefore \quad t = \frac{\Delta v}{a}$$

where $\Delta v = 170\,\text{km/h} - 100\,\text{km/h} = 70\,\text{km/h} = \dfrac{70\,\text{km/h} \times 1000\,\text{m/km}}{3600\,\text{s/h}}$

$$= 19.4\,\text{m/s}$$

and $\quad a = 1.8\,\text{m/s}^2$

$$\therefore \quad t = \frac{19.4\,\text{m/s}}{1.8\,\text{m/s}^2}$$

$$= 10.8\,\text{s}$$

i.e. the time required to reduce the speed to 100 km/h is 10.8 s.

6.6 Velocity–time graphs

Example 1 In a laboratory experiment on a trolley moving in a straight line, the following values for velocity and time were obtained:

Time t (s)	0	1	2	3	4	5
Velocity v (mm/s)	0	5	10	15	15	15

Plot the graph of velocity against time and use it to determine (a) the acceleration of the trolley, (b) the distance moved by the trolley in 5 seconds, (c) the average velocity of the trolley.

a) Referring to the velocity–time graph in fig. 6.8, between O and A the graph is a straight line; i.e. between O and A, the *change* in velocity, Δv, is directly proportional to the time. This may be written mathematically in the form

$\Delta v \propto t$ (the symbol $\propto$ means '*proportional to*')

$$\therefore \quad \Delta v = at$$

The constant a is the acceleration,

$$\therefore \quad a = \frac{\Delta v}{t}$$

i.e. *the slope or gradient of a velocity–time graph represents the acceleration, which should be remembered.*

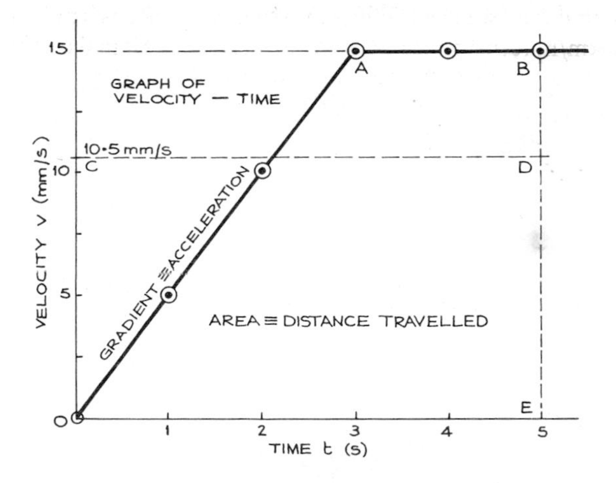

Fig. 6.8 Velocity–time graph for an experiment on a moving trolley

From the graph,

$$\text{slope of OA} = \frac{15\,\text{mm/s}}{3\,\text{s}}$$

$$= 5\,\text{mm/s}^2$$

i.e. the acceleration of the trolley is $5\,\text{mm/s}^2$.

b) Average velocity $= \dfrac{\text{distance travelled in a specified direction}}{\text{time taken}}$

$\therefore$ distance travelled = average velocity × time

i.e. the area of the velocity–time graph represents the distance travelled, which should be remembered.

$\therefore$ Distance travelled $\equiv$ area of velocity–time graph

$$= \tfrac{1}{2} \times 3\,\text{s} \times 15\,\text{mm/s} + 15\,\text{mm/s} \times 2\,\text{s}$$

$$= 52.5\,\text{mm}$$

i.e. the distance travelled is 52.5 mm.

c) Referring to the graph in fig. 6.8, the total area under the line OAB is equal to the area of the rectangle OCDE,

where OC = DE = average velocity and CD = 5 s = time

$\therefore$ distance travelled = average velocity × time

which it is useful to remember.

$$\therefore \quad \text{Average velocity} = \frac{\text{distance travelled}}{\text{time taken}}$$

$$= \frac{52.5\,\text{mm}}{5\,\text{s}}$$

$$= 10.5\,\text{mm/s}$$

i.e. the average velocity is 10.5 mm/s.

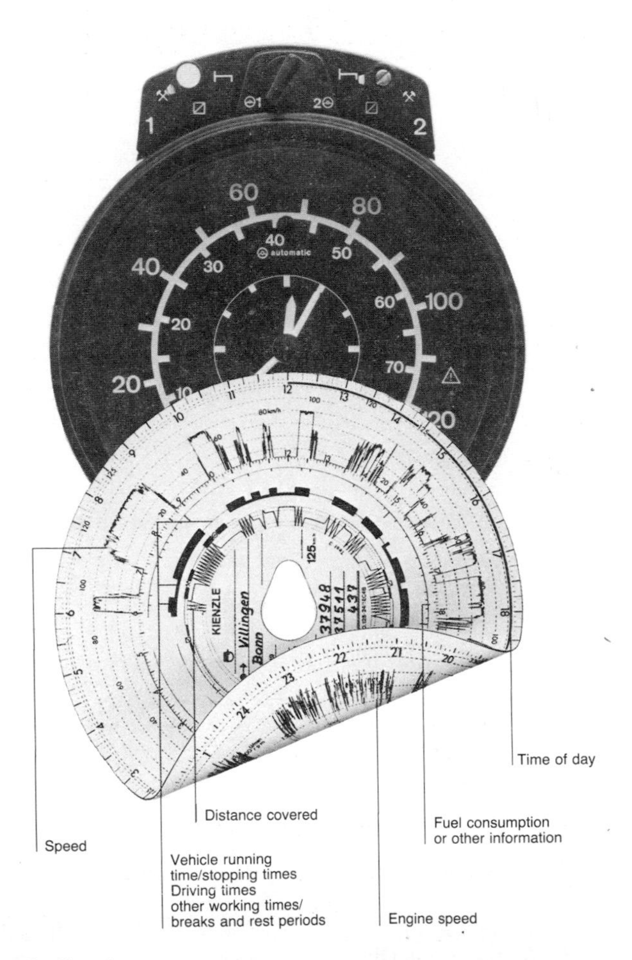

A tachograph, fitted to heavy-goods vehicles and long-distance coaches to record data on distance travelled, fuel consumption, and engine speed

Example 2 A motor car accelerates uniformly from rest for 10 seconds until its velocity is 20 m/s. The car maintains this speed for 90 seconds and then slows uniformly to rest in 2 seconds. Sketch the velocity-time graph and use it to find (a) the acceleration and retardation, (b) the total distance travelled, (c) the average velocity.

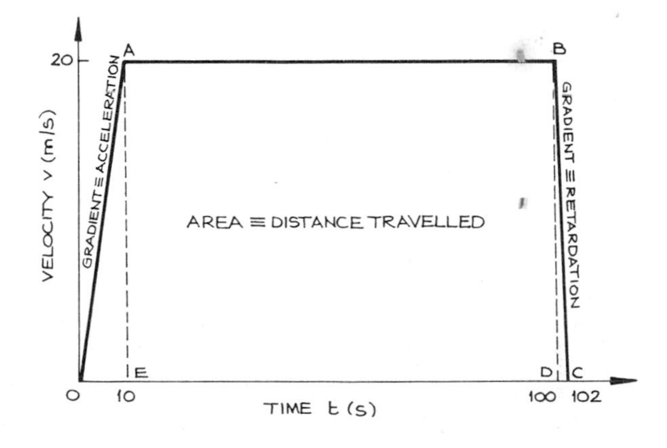

Fig. 6.9

a) Referring to the velocity-time graph in fig. 6.9, the acceleration of the motor car is represented by the gradient of the line OA, while the retardation is represented by the gradient of the line BC;

$\therefore$ acceleration $\equiv$ gradient of OA

$$= \frac{20\,\text{m/s}}{10\,\text{s}} = 2\,\text{m/s}^2$$

and retardation $\equiv$ gradient of BC

$$= -\frac{20\,\text{m/s}}{2\,\text{s}} = -10\,\text{m/s}^2$$

(the minus sign indicates that the motor car is slowing down or retarding). i.e. the acceleration and retardation of the motor car are $2\,\text{m/s}^2$ and $10\,\text{m/s}^2$ respectively.

b) Total distance travelled $\equiv$ area of velocity-time graph

$$\equiv \text{area (OAE + ABDE + BCD)}$$

$$= \tfrac{1}{2} \times 20\,\text{m/s} \times 10\,\text{s} + 20\,\text{m/s} \times 90\,\text{s} + \tfrac{1}{2} \times 20\,\text{m/s} \times 2\,\text{s}$$

$$= 1920\,\text{m} \quad \text{or} \quad 1.92\,\text{km}$$

i.e. the total distance travelled is 1.92 km.

c) Average velocity $= \dfrac{\text{distance travelled}}{\text{time taken}}$

where time taken $= 10\,\text{s} + 90\,\text{s} + 2\,\text{s} = 102\,\text{s}$

$\therefore$ average velocity $= \dfrac{1920\,\text{m}}{102\,\text{s}}$

$= 18.82\,\text{m/s}$

i.e. the average velocity is 18.82 m/s.

6.7 Force and acceleration

The wheeled trolley shown in fig. 6.10 is in equilibrium on a smooth and horizontal surface. Attached to the trolley is a length of string which passes over the pulley such that, when the string is pulled downwards, the trolley will move towards the pulley.

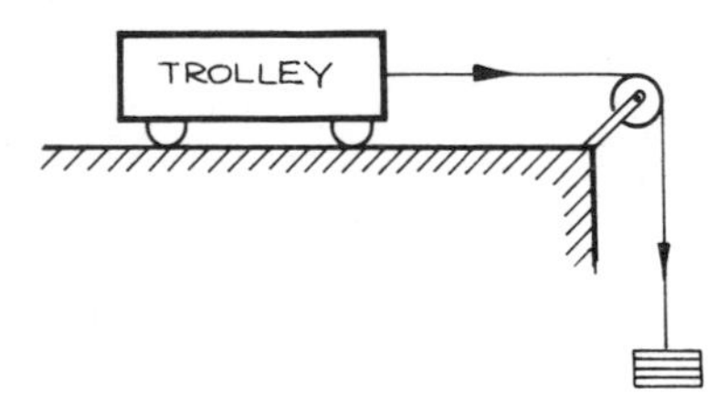

Fig. 6.10

With the trolley at rest, the pull in the string may be gradually increased by attaching small loads to the free end of the string until the trolley just starts to move. If no further loads are added, it will be observed that the trolley is moving at constant velocity.

If the load on the string is increased further, it will be observed that the velocity of the trolley will increase steadily - i.e. the trolley will be accelerating. Further experimentation will reveal that, as the load on the string is increased further, the acceleration will also increase.

With the trolley accelerating,

$$\text{total force acting on the trolley} = \text{force to overcome resistance to motion} + \text{force to accelerate the trolley}$$

or $$\text{accelerating force} = \text{total force} - \text{force to overcome resistance to motion}$$

which is the *net* force acting on the trolley.

i.e., when a force is applied to a body causing it to move, the *net force* acting will accelerate the body in the direction of the force, *which should be remembered.*

The initial force which is resisting motion is known as the *frictional resisting force* or simply the *friction force*. It should be noted that friction

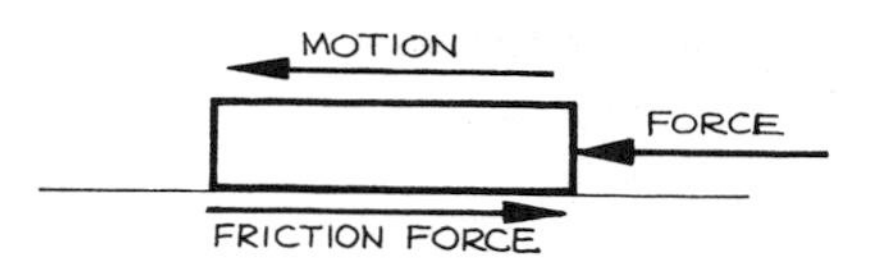

Fig. 6.11 Friction force

forces are *always* present between bodies which are in sliding contact and they act in the opposite direction to the motion. This is illustrated in fig. 6.11.

6.8 Freely falling bodies

When a body is falling freely towards the earth, it is being attracted towards the centre of the earth by the *force due to gravity* or *gravitional force.* Experiments conducted in a vacuum have shown that *any* falling object, be it a feather or a 10 tonne tank, accelerates toward the surface of the earth at a constant rate. This acceleration is called the *acceleration due to gravity* and is denoted by the symbol g.

On the earth, the magnitude of the acceleration due to gravity g varies from 9.78 m/s^2 at the equator to 9.832 m/s^2 at the North and South Poles. In London at sea level, $g = 9.80665\,\text{m/s}^2$. On the moon a freely falling body would experience a gravitational acceleration of approximately 2 m/s^2 towards the centre of the moon.

When bodies fall freely in the atmosphere, their downward motion is resisted by the air. How the air resistance affects the downward acceleration depends upon the shape and size of the body. For example, a 'free-fall' parachutist who jumps from an aircraft and remains in the vertical position

Free-fall parachutists in conference! The positions of their arms and legs control the parachutist's rate of descent – if they are tucked in, they will fall more quickly

will fall more quickly than a parachutist who is falling in the horizontal position. This is because the parachutist in the horizontal position presents a greater surface area to the air.

When the parachute for each case above opens, the parachutist will descend more slowly, since the upward resisting force of the air on the parachute canopy considerably reduces the effect of the downward gravitational force on the parachutist.

To sum up, in the absence of any upward resisting force such as air resistance, a freely falling body will accelerate downwards at a constant rate which, for calculation purposes, is taken to be 9.81 m/s².

Example A man working at the top of a ladder drops a hammer which falls freely to the ground 12 m below. Assuming the initial velocity of the hammer was zero, and neglecting air resistance, calculate the velocity of the hammer as it strikes the ground. Find also the time taken for the hammer to fall 12 m.

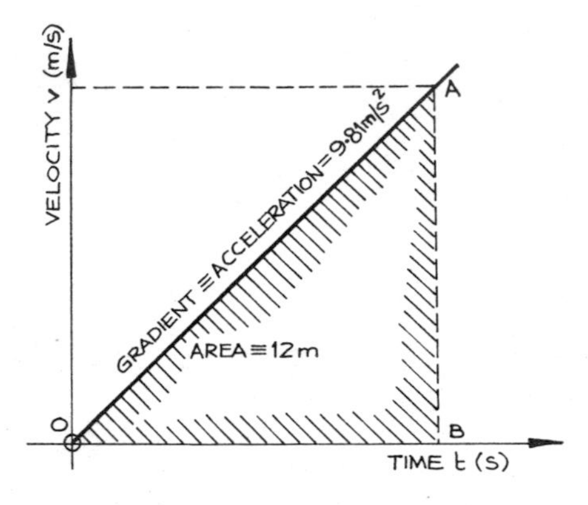

Fig. 6.12

Referring to the velocity–time graph shown in fig. 6.12, let v be the velocity at time t after the hammer has fallen 12 m. In fig. 6.12,

$$\text{distance travelled} \equiv \text{area OAB} = \tfrac{1}{2}vt = 12\,\text{m}$$

$$\therefore \qquad t = \frac{2 \times 12\,\text{m}}{v} = \frac{24\,\text{m}}{v} \qquad \text{(i)}$$

Also, acceleration $\equiv$ slope or gradient of OA $= 9.81\,\text{m/s}^2 = v/t$

$$\therefore \quad t = \frac{v}{9.81\,\text{m/s}^2} \qquad \text{(ii)}$$

Now equation (i) = equation (ii)

$$\therefore \quad \frac{24\,\text{m}}{v} = \frac{v}{9.81\,\text{m/s}^2}$$

$$\therefore \quad v^2 = 9.81\,m/s^2 \times 24\,m$$

$$= 235.44\,(m/s)^2$$

$$\therefore \quad v = 15.34\,m/s$$

i.e. the velocity of the hammer as it strikes the ground is 15.34 m/s.

To find the time taken to fall 12 m, substitute $v = 15.34$ m/s in either equation (i) or equation (ii). Substituting in equation (ii),

$$t = \frac{15.34\,m/s}{9.81\,m/s^2}$$

$$= 1.56\,s$$

i.e. the hammer takes 1.56 s to fall through a distance of 12 m.

Exercises on chapter 6

1 Convert the following into metres per second (m/s): (a) 36 km/h, (b) 48 km/h, (c) 15 km/h, (d) 72 km/h, (e) 100 km/h, (f) 240 m/min, (g) 600 mm/s.

2 Convert the following into kilometres per hour (km/h): (a) 3 m/s, (b) 24 m/s, (c) 80 m/min, (d) 240 mm/s.

3 A man walks from point A to point B as follows: from A, 200 m due north then 100 m due west then 50 m due north-west to B. Find (a) the distance travelled, (b) the displacement between A and B. If the time taken is 1 min 55 s, find the average speed and velocity of the man.

4 Define *speed*. An aircraft flies due west for 2 hours at an average speed of 800 km/h followed by 1.5 hours at 750 km/h in a north-westerly direction. Find the distance travelled and the displacement of the aircraft.

5 In a cycle race over 100 km, the average speed of the winner was 40 km/h while his average velocity was 8 m/s. Find the minimum distance between the start and finish of the race in kilometres.

6 On a 400 m running track, the start and finish points are at the same place and half-way along the straight which is 100 m long. In a 400 m sprint race, runner A is in the inside lane while runner B is in the outside lane and, so that both runners cover the same distance, runner B is positioned 49.76 m in front of runner A at the start of the race. In the race, A and B tie for first place in a time of 48 s. Find the average velocity and speed of each runner.

7 A motor car starting from rest accelerates uniformly for 15 s during which time the distance travelled is 600 m. Find (a) the acceleration; (b) the speed of the car after (i) 9 s, (ii) 15 s.

8 Using the following data, draw a distance–time graph and use it to determine the average speed between (a) $t = 0$ and $t = 20$ s, (b) $t = 0$ and $t = 50$ s.

Distance travelled (km)	0.5	1.2	1.9	2.8	3.4
Time taken (s)	10	20	30	40	50

9 The following velocity–time data were obtained during a study of an article on a conveyor mechanism:

0 to 1 min – uniform increase in velocity from 0 to 0.2 m/s;
1 to 3 min – velocity remains constant at 0.2 m/s;
3 to 4 min – uniform increase in velocity from 0.2 to 0.6 m/s;
4 to 5 min – velocity remains constant at 0.6 m/s;
5 to 8 min – velocity decreases uniformly from 0.6 m/s to zero.

Sketch the velocity– time graph and use it to find (a) the distance travelled by the article, (b) the acceleration of the article between $t = 0$ and $t = 1$ min and between $t = 3$ min and $t = 4$ min, (c) the deceleration or retardation between $t = 5$ min and $t = 8$ min.

10 A stone is dropped with zero initial velocity down a disused water well, reaching the bottom in 2 seconds. Find the depth of the well and the maximum velocity of the stone.

11 The velocity–time graph shown in fig. 6.13 describes the motion of a two-speed electric vehicle. Find (a) the distance travelled in 115 s, (b) the maximum acceleration, (c) the retardation.

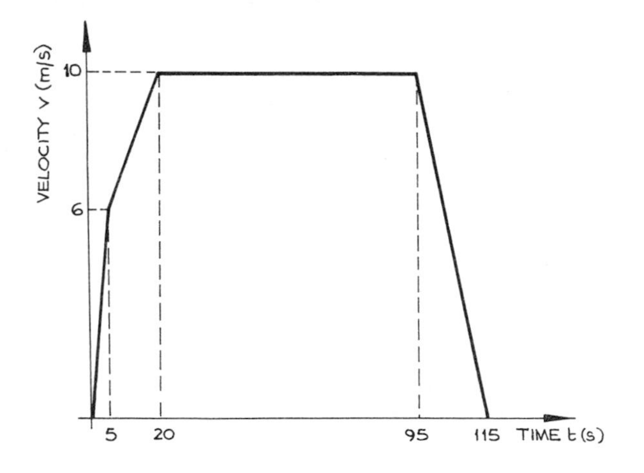

Fig. 6.13

12 A motor car, starting from rest, accelerates uniformly for 5 s until it is travelling at 45 km/h. The vehicle maintains this speed for 30 s before braking and coming uniformly to rest in 2 s. Find the acceleration and retardation of the vehicle and the total distance travelled.

13 A sledge is towed along a snow-covered footpath. With the aid of a neat sketch, show, relative to the motion of the sledge, the direction of all the forces acting on it.

14 In an emergency air-lift to a remote part of Africa, some packages are dropped from the aircraft with a parachute and some are dropped without. Which packages will fall more quickly to the ground and why?

15 A man dives from a platform into a pool 30 m below him. If the initial velocity of the man is zero, calculate the time taken and the velocity of the diver as he reaches the water.

16 From London to Manchester by road is 338 km. Measured on a map, the shortest distance between the cities is 274 km. An express motor coach leaves London at 3.30 p.m. and arrives in Manchester at 7.10 p.m. Find (a) the time for the journey, (b) the average speed, (c) the average velocity.

7 Waves

7.1 Introduction

The waves with which we are probably most familiar are water waves, produced by some disturbance in the water, such as a stone dropped into the water or a boat passing, or by wind disturbance. The shape of water waves is typical of the shape of all types of wave in that they have *crests* and *troughs* and they have a *regularly occurring waveform*.

Other examples of wave motion are light waves, radio waves, and sound waves.

7.2 Wave properties

As we shall see, sound waves are fundamentally different from light waves and radio waves, but there are some properties which are common to waves of any type.

Firstly, waves *travel* - that is, they move away from the source which is producing them with a certain velocity.

Light waves move in free space (i.e. in a vacuum) with the extremely high velocity of 3×10^8 m/s (which is equivalent to 670000000 miles per hour). Sound waves are much more sedate - their velocity in normal air is 330 m/s (equivalent to 740 miles per hour).

The velocity of any wave depends on the material through which the wave travels. Light waves travel more slowly in water than they do in air, whereas sound waves travel more rapidly in water than they do in air.

The second property common to all waves is their *regularity* - that is, crest follows trough and trough follows crest for as long as the waves are produced. In this respect they may be likened to bullets appearing from the muzzle of a machine gun at a regular rate. We may consider the rate at which the

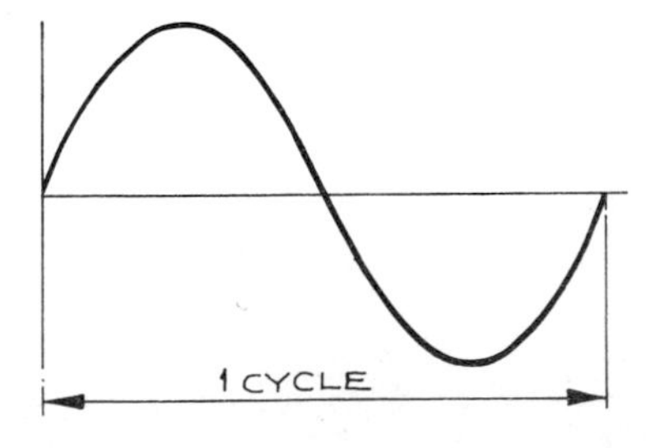

Fig. 7.1 One cycle of a wave

bullets appear from the gun to be the frequency. Similarly with waves - the number of crests (or troughs) that passes a fixed point in a second is referred to as the *frequency* of the waves. One full *cycle* of a wave is shown in fig. 7.1.

Frequency (symbol f) is measured in hertz (symbol Hz), which is the same as the number of cycles per second. (The unit of frequency is named after the German scientist Heinrich Hertz (1857-1894) who first discovered the existence of radio waves.)

The distance between successive wave crests is referred to as the *wavelength*. This may be likened to the distance that a bullet has travelled from the gun before another bullet appears. This of course depends on the velocity of the bullets and on the frequency with which they occur. Wavelength is given the symbol λ (*lambda*) and is measured in metres.

The relationship between wave velocity v, frequency f, and wavelength λ is

$$v = f\lambda$$

We will demonstrate this by means of an example.

Example 1 A boat sailing up a river produces waves which travel away from the boat at 0.4 m/s. The waves travel past a fixed post in the stream at two every second as shown in fig. 7.2. Calculate the distance between successive crests.

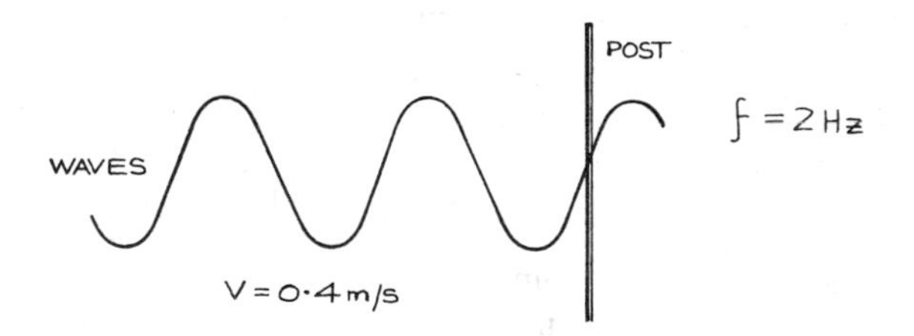

Fig. 7.2 Water waves travelling past a fixed post

Since two crests pass the post each second, then the frequency f is 2 Hz, or two cycles each second.

Since the wave velocity v is 0.4 m/s, then the distance travelled by the wave is 0.4 m in 1 second.

The distance between successive crests must therefore be 0.4 m/2 = 0.2 m (i.e. the wavelength λ is 0.2 m).

It is evident from example 1 that the wavelength is given by

$$\lambda = \frac{v}{f}$$

Rearranging this equation gives the relationship which is true for all waves:

$$v = f\lambda$$

i.e. the wave velocity in metres per second is equal to the frequency in hertz multiplied by the wavelength in metres.

Example 2 BBC Radio 4 transmits at a wavelength of 1500 m. Radio waves travel at about 3×10^8 m/s. Calculate the transmission frequency.

$$v = f\lambda$$

$$\therefore \quad f = \frac{v}{\lambda}$$

where $v = 3 \times 10^8$ m/s and $\lambda = 1500$ m

$$\therefore \quad f = \frac{3 \times 10^8 \text{ m/s}}{1500 \text{ m}}$$

$$= 200000 \text{ Hz} \quad \text{or} \quad 200 \text{ kHz}$$

i.e. the transmission frequency is 200 kHz (waves about 1 mile long are passing at the rate of 200000 every second).

Example 3 A local radio station transmits on VHF (very high frequency) at 92.4 MHz. Calculate the wavelength of the transmission.

$$\lambda = \frac{v}{f}$$

where $v = 3 \times 10^8$ m/s and $f = 92.4 \text{ MHz} = 92.4 \times 10^6$ Hz

$$\therefore \quad \lambda = \frac{3 \times 10^8 \text{ m/s}}{92.4 \times 10^6 \text{ Hz}}$$

$$= 3.25 \text{ m}$$

i.e. the wavelength is 3.25 m.

Example 4 An ultrasonic transducer (i.e. a device capable of producing sound at a frequency above the audible range) produces a signal at 40 kHz. If the wavelength is 8.25 mm, calculate the speed of sound.

$$v = f\lambda$$

where $f = 40 \text{ kHz} = 40 \times 10^3$ Hz

and $\lambda = 8.25 \text{ mm} = 8.25 \times 10^{-3}$ m

$$\therefore \quad v = 40 \times 10^3 \text{ Hz} \times 8.25 \times 10^{-3} \text{ m}$$

$$= 330 \text{ m/s}$$

i.e. the velocity of sound is 330 m/s.

7.3 Transverse and longitudinal waves

When attempting to investigate 'what waves are', it is best to consider wave motions with which we are familiar. Water waves are very convenient since they behave in many respects like light waves and radio waves.

We have said that waves travel, and the example of a water wave travelling past a fixed post demonstrates this. However, a cork floating on the surface of the water moves up and down with the wave motion but does not travel along with the wave.

This illustrates a property which water waves have in common with light and radio waves - the displacement of the wave (i.e. the vibration) is at right angles to the direction of travel. This type of wave is called a *transverse* wave, since the displacement is across the direction of travel. It is the most obvious type of wave motion, and has a wavy shape to its motion as shown in fig. 7.3.

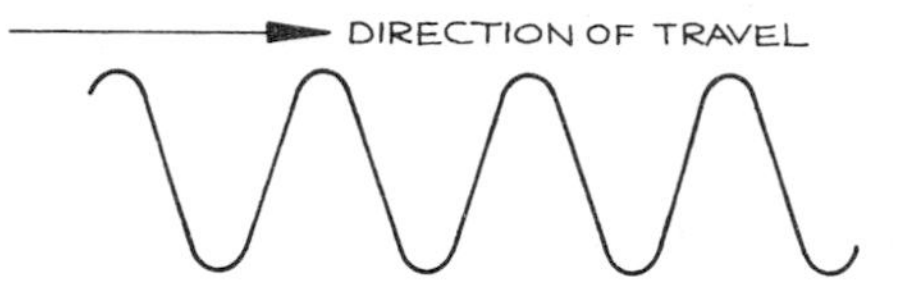

Fig. 7.3 Wave motion typical of transverse waves

Another type of wave motion is that with which sound travels. In this case the wave transmission is due to alternating high and low pressure - or compressions and rarefactions - in the air (i.e. the medium in which the sound travels).

These compressions and rarefactions are in the same direction as the direction of travel of the wave. This does not mean that the air travels from the sound source to the listener, but rather that air molecules vibrating back and forth about a fixed position, in the direction of travel of the wave, disturb adjacent air molecules and thus the sound progresses as shown in fig. 7.4.

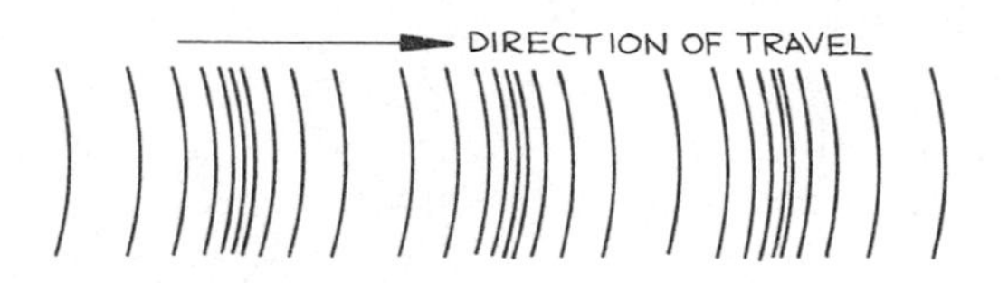

Fig. 7.4 Compressions and rarefactions typical of sound-wave motion

This type of wave is called a *longitudinal* wave, since the disturbance is in the same direction as the direction of travel of the wave.

A good visual demonstration of longitudinal motion is possible using a 'slinky'. This is a long coiled spring which is very 'sloppy'. Laying the spring slightly stretched on a table and then hitting one end of the spring demonstrates how the compression moves through the whole length of the spring, and this is typical of the longitudinal wave motion of sound.

Example A rope is fixed at one end and waggled up and down at the other end so that a continuous train of waves moves along the rope. Is this wave motion transverse or longitudinal?

Transverse. The motion is at right angles to the direction of travel.

7.4 Electromagnetic radiation

Light waves and radio waves are both part of the same type of radiation called *electromagnetic radiation*. This can travel through a vacuum and therefore does not require a transmission medium, which is why we can receive light and radio waves from distant stars through the vacuum of space. The whole range of electromagnetic radiation is known as the *electromagnetic spectrum*.

The radio waves which are used to carry the various radio and television channels are broadcast from high-power transmitters which are situated in high places or on towers so that signal reception will be possible over a wide area. We receive these waves on radio or television sets by means of an aerial

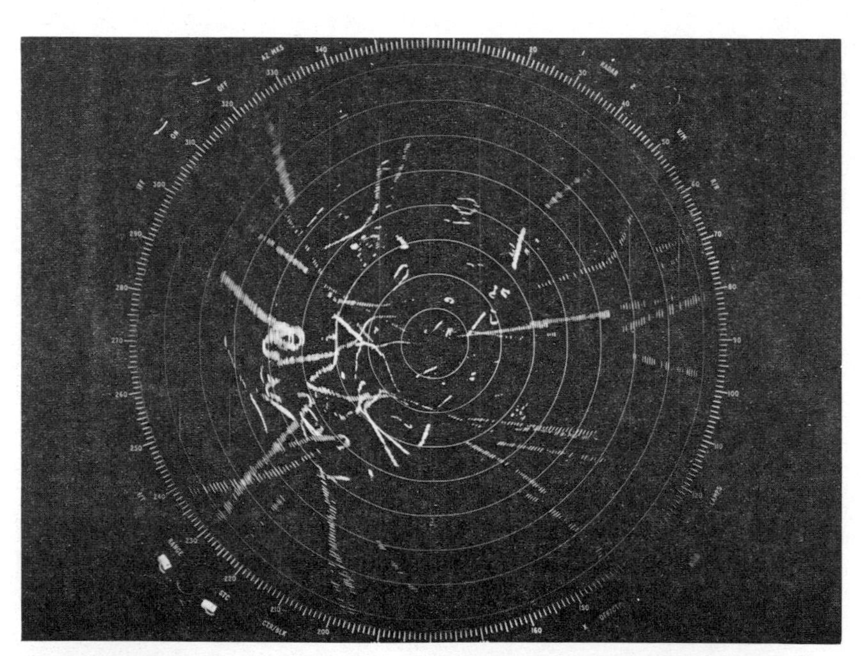

The picture built up on the screen of a cathode-ray tube by a rotating radar scanner

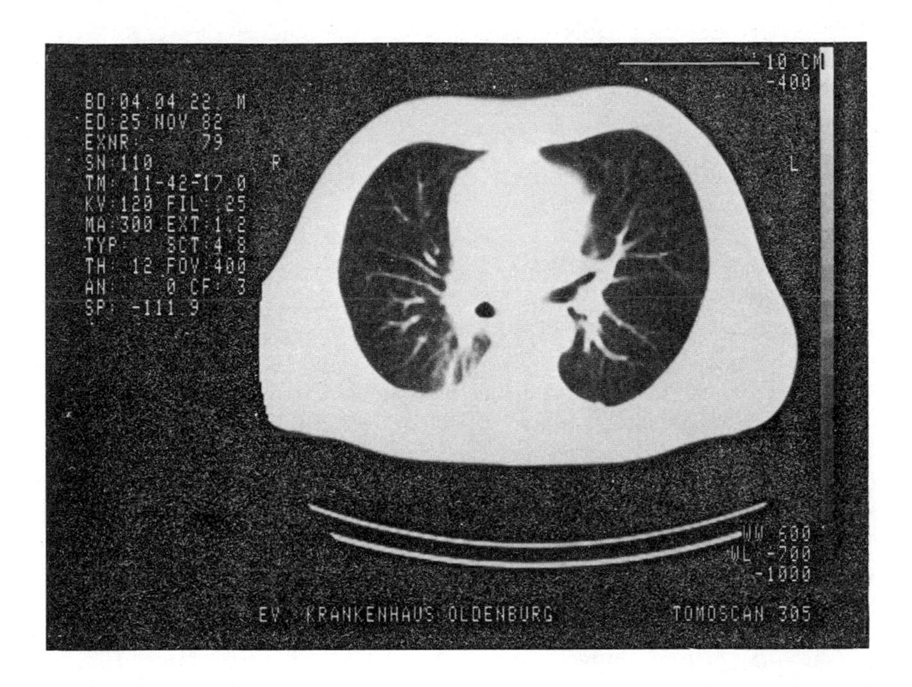

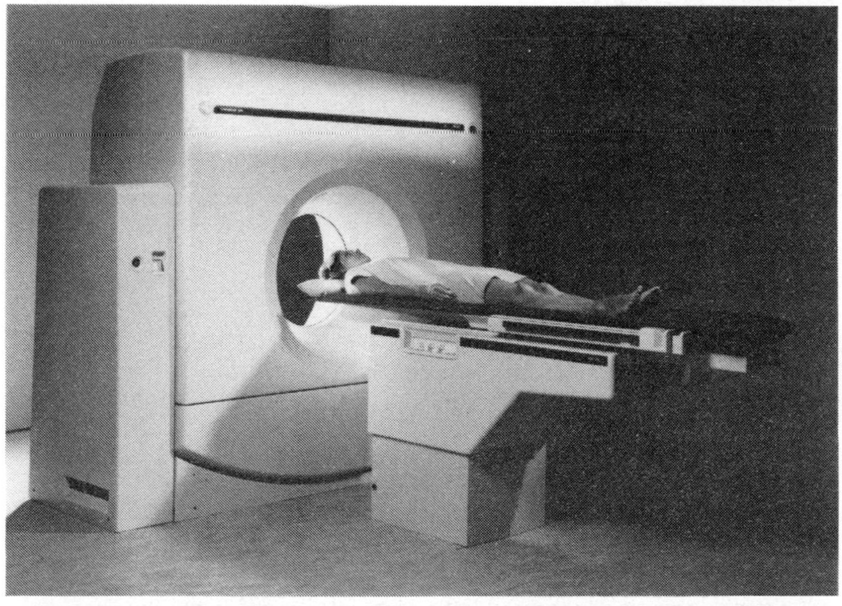

An image of the lungs, produced by the computerised tomography scanner shown below, which uses X-rays to provide a general-purpose body scanner

connected to a receiver which has a circuit that, by 'tuning', enables individual waves to be selected from all those being broadcast.

Radar uses radio waves to detect the position of ships, aircraft, and land masses by transmitting radio pulses and detecting the time taken until the reflected signal is received. When this is done continually through a full circle, using a rotating radar scanner, a complete picture of the surroundings is built up.

Another example of electromagnetic radiation is heat rays or infra-red radiation, such as that produced by a coal fire or a radiator. These rays are evidenced by the fact that they keep us warm, although the infra-red radiation cannot be seen by the eye – we 'see' a glowing coal fire as a result of it also emitting light waves.

Infra-red radiation is used in medicine to record the heat pattern of the body, and in burglar alarms which can detect the body heat of the potential burglar.

Yet another type of electromagnetic radiation is ultra-violet radiation, which has a wavelength just below that of visible light. It is used in ultraviolet recorders (which are used to record electrical waveforms) and it affects photographic film. It also produces a nice tan when the body has been exposed to it for some time, and can cause sunburn.

X-rays are part of the electromagnetic spectrum and are used in medicine to take X-ray pictures through the body, showing the bone structure. They are used in a similar way in industrial situations to 'see' the internal construction of objects.

7.5 Sound waves

Sound waves need a medium in which to travel, say air or water. Sound waves will not travel in a vacuum.

The frequency range of sound waves audible to the human ear is from about 20 Hz (20 cycles per second) up to about 15 kHz (15000 cycles per second). Above this frequency, sound waves are referred to to as ultrasonic, with the lower limit being audible to some animals and the upper limit being determined by the practical means of producing it.

Ultrasonic waves have a number of useful applications. They are used in cleaning and in welding, since they produce a high-frequency vibration. (In this method of welding, the two materials are 'scrubbed' together at high frequency, which helps to form a mechanical bond.)

They are also used in echo-sounding, particularly under water, where the electrical conductivity of water makes electromagnetic (radio) waves of comparable wavelength suffer from severe attenuation (loss of signal). Ultrasonic waves are emitted from a ship towards the sea bottom, and the time taken to receive the reflected signal is a measure of the distance to the bottom. In this application, ultrasonic waves are referred to as 'sonar'.

They are used in a similar way to detect cracks in metal castings, and bats use ultrasonic waves as a form of 'radar' for detecting objects in their flight path.

They are also used in medicine to display the internal organs of the body and to monitor the growth of unborn babies, and in these applications they are referred to as 'ultrasound'.

Exercises on chapter 7

1 (a) Define *wave frequency*. (b) A tuning fork vibrating at a frequency of 248 Hz produces a sound wave of wavelength 1.166 m. Calculate the speed of sound in air.

2 (a) Give two examples of wave motion. (b) Sketch the pressure wave shown in fig. 7.5 and mark on it the wavelength and one cycle.

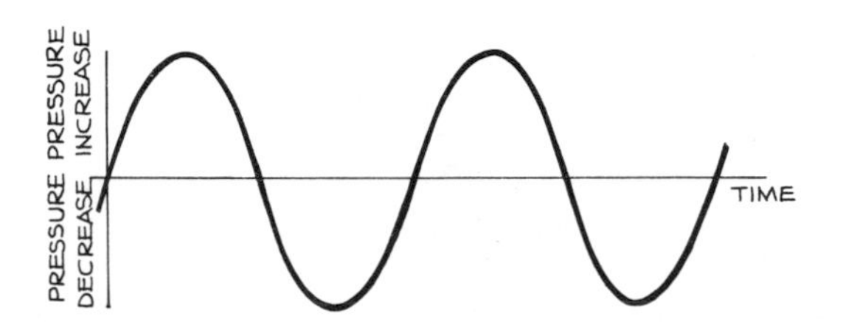

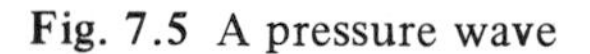

Fig. 7.5 A pressure wave

3 State how long it would take for a sound wave of frequency 10 kHz and wavelength 33 mm to travel a distance of 500 m in air.

4 (a) State the equation for wave motion which relates velocity v, frequency f, and wavelength λ. (b) Light travels in air with a velocity of 3×10^8 m/s. Calculate the frequency of red light which has a wavelength of 0.7 μm.

5 With the aid of a simple diagram, explain the meaning of wavelength and frequency.

6 (a) Explain how sound can travel through air but not through a vacuum. (b) Calculate the wavelength in both air and water of a sound of frequency 1 kHz if the speed of sound in air is 335 m/s and in water is 1360 m/s.

7 BBC Radios 3 and 4 are broadcast on VHF/FM at between 90.2 to 92.4 MHz and 92.4 to 94.7 MHz respectively, the exact frequency depending on the area. Calculate the corresponding wavelengths. Radio waves travel at 3×10^8 m/s.

8 Infra-red radiation is in the wavelength band 1 mm to 1 μm. Calculate the corresponding frequencies. Light waves travel at 3×10^8 m/s.

9 (a) State the difference between transverse and longitudinal waves. (b) Water waves in a tank may be used to demonstrate some of the aspects of motion of what type of waves?

8 Work, energy, and power

8.1 Work and energy

In physical science, *work* is defined as the force required to overcome resistance to motion multiplied by the distance moved in the direction of the force. *Energy* is the *capacity* or *ability* to do work.

Examples of work and energy include:

a) a lift carrying passengers from the ground floor to the top of a building - work is done in overcoming the force due to gravity, while the energy is drawn from the electricity supply;
b) the body-builder using a spring-type chest-expander - work is done in overcoming the stiffness of the spring, the energy being provided by the food the body-builder has eaten.
c) a tractor hauling a trailer - work is done in overcoming frictional resistance to motion, and the energy is provided by the petrol or diesel fuel.

Thus work = force to overcome resistance to motion × distance moved in the direction of the force

or $W = Fs$

which should be remembered.

The unit for work is the joule (symbol J). The joule is defined as the work done when the point of application of a force of one newton is displaced through a distance of one metre in the direction of the force;

i.e. $1\,\text{J} = 1\,\text{N} \times 1\,\text{m}$

or $1\,\text{J} = 1\,\text{Nm}$

which should be remembered.

It should be noted that the joule is also the unit of energy.

If a force is applied and there is no motion, then there is no work done, *no matter how large the force.*

Example 1 A force of 120 N is applied to a crate, causing it to slide for a distance of 5 m in the direction of the force. Calculate the work done on the crate.

$W = Fs$

where $F = 120\,\text{N}$ and $s = 5\,\text{m}$

$\therefore \quad W = 120\,\text{N} \times 5\,\text{m}$

$= 600\,\text{J} \quad (\text{since } 1\,\text{J} = 1\,\text{Nm})$

i.e. the work done on the crate is 600 J.

In the tug of war shown, (a) the rope is being stretched and (within the elastic limit) will obey Hooke's law; (b) at the start of the pull there is no movement and the teams are in equilibrium – no work is done no matter how hard they are pulling; (c) as the pull proceeds, the winning team does work on the losing team.

Example 2 The work done on a body in moving it through a distance of 15 m is 500 J. Find the applied force in the direction of the motion.

$$W = Fs$$

$$\therefore \quad F = \frac{W}{s}$$

where $W = 500\,\text{J}$ and $s = 15\,\text{m}$

$$\therefore \quad F = \frac{500\,\text{J}}{15\,\text{m}}$$

$$= 33.3\,\text{N}$$

i.e. the force on the body is 33.3 N.

Example 3 The work done on a body by a force of 60 N is 780 J. Find the distance moved by the body in the direction of the force.

$$W = Fs$$

$$\therefore \quad s = \frac{W}{F}$$

where $W = 780\,\text{J}$ and $F = 60\,\text{N}$

$$\therefore \quad s = \frac{780\,\text{J}}{60\,\text{N}}$$

$$= 13\,\text{m}$$

i.e. the distance moved is 13 m.

8.2 Work diagrams

A work diagram is a graph of force against distance moved by the force. The area of a work diagram represents the work done by the force.

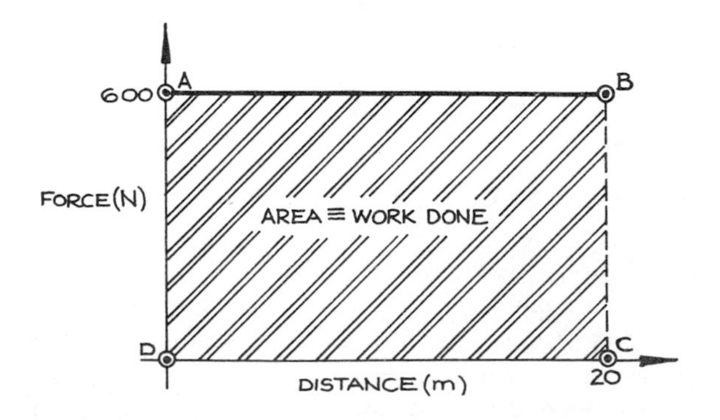

Fig. 8.1 Work diagram for constant force

Figure 8.1 shows a work diagram for a force of 60 N acting through a distance of 20 m.

$$\text{Work done} \equiv \text{area ABCD}$$
$$= 600\,\text{N} \times 20\,\text{m}$$
$$= 12000\,\text{J}$$

Work diagrams are used to determine the work done by *variable* forces.

Example 1 In a test on a compression spring, the following data were obtained:

Force (N)	0	80	160	240	320
Compression (mm)	0	10	20	30	40

Plot a graph of force against compression and thus determine the total work done in compressing the spring 40 mm.

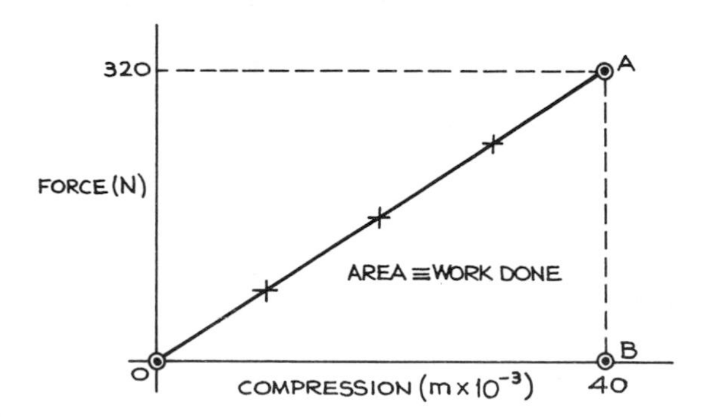

Fig. 8.2 Force–extension graph for a spring

From the data, the graph shown in fig. 8.2 may be plotted.

$$\text{Work done} \equiv \text{area OAB}$$
$$= \tfrac{1}{2} \times 320\,\text{N} \times 0.04\,\text{m}$$
$$= 6.4\,\text{J}$$

i.e. the work done in compressing the spring is 6.4 J.

Referring to example 1,

$$\begin{array}{l}\text{work done in compressing}\\ \text{(or extending) a spring}\end{array} = \tfrac{1}{2} \times \text{force} \times \begin{array}{l}\text{distance spring is}\\ \text{compressed (or extended)}\end{array}$$

or

$$\text{work done} = \text{average force} \times \begin{array}{l}\text{compression or}\\ \text{extension of spring}\end{array}$$

which it is useful to remember.

Example 2 A truck was moved through a distance of 100 m. For the first 10 m of motion, the force acting on the truck was a constant 200 N. During the next 50 m, the force decreased uniformly at the rate of 2 N/m, the remaining distance being covered with the force remaining constant. Draw the work diagram and determine the total work done on the truck.

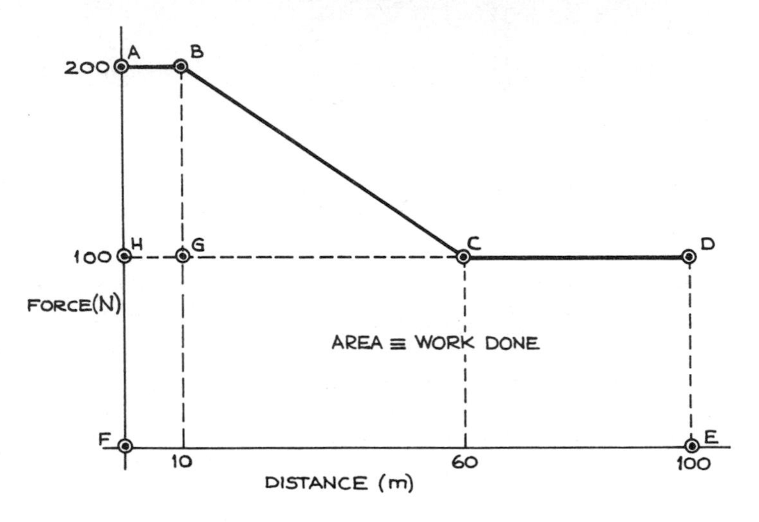

Fig. 8.3 Work diagram

From the work diagram in fig. 8.3,

$$\begin{aligned} \text{work done} &\equiv \text{area ABCDEF} \\ &\equiv \text{area ABGH} + \text{area BCG} + \text{area HDEF} \\ &= (100\,\text{N} \times 10\,\text{m}) + \tfrac{1}{2}(100\,\text{N} \times 50\,\text{m}) + (100\,\text{N} \times 100\,\text{m}) \\ &= 1000\,\text{Nm} + 2500\,\text{Nm} + 10000\,\text{Nm} \\ &= 13500\,\text{J} \end{aligned}$$

i.e. the total work done is 13500 J.

Example 3 The force–distance graph shown in fig. 8.4 was obtained during a test on a trailer linkage mechanism. Determine the total work done.

From the vertical and horizontal scales in fig. 8.4,

$$1\,\text{mm}^2 \equiv 10\,\text{N} \times 2\,\text{m} = 20\,\text{J}$$

$$\begin{aligned} \therefore\ \text{work done} &= \text{area ABCD} \times \text{scale} \\ &= 672\,\text{mm}^2 \times 20\,\text{J/mm}^2 \\ &= 13\,440\,\text{J} \quad \text{or} \quad 13.44\,\text{kJ} \end{aligned}$$

i.e. the total work done is 13.44 kJ.

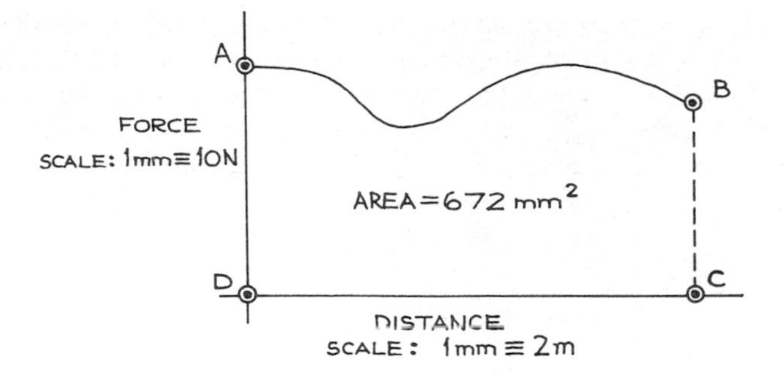

Fig. 8.4 Force–distance graph for a trailer linkage mechanism

Example 4 The force-distance graph shown in fig. 8.5 was obtained from a test on a wheeled vehicle. Determine the total work done on the vehicle.

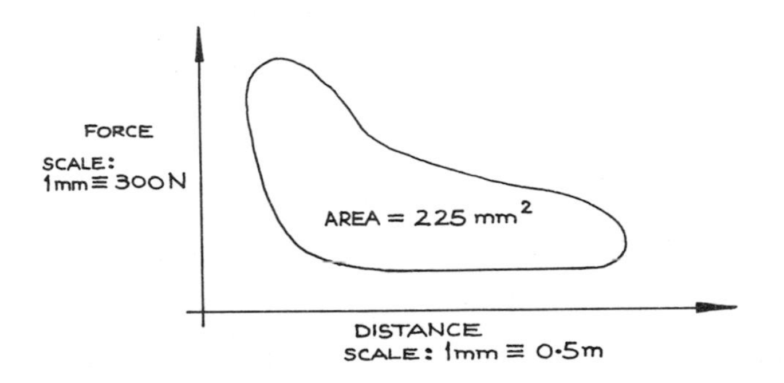

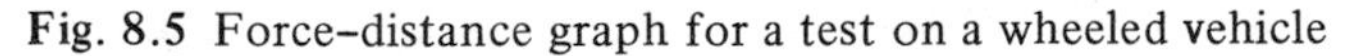

Fig. 8.5 Force–distance graph for a test on a wheeled vehicle

From the vertical and horizontal scales in fig. 8.5,

$$1\ \text{mm}^2 \equiv 300\ \text{N} \times 0.5\ \text{m} = 150\ \text{J}$$

$$\therefore \text{ work done} = \text{area of work diagram} \times \text{scale}$$

$$= 225\ \text{mm}^2 \times 150\ \text{J/mm}^2$$

$$= 33\,750\ \text{J} \quad \text{or} \quad 33.75\ \text{kJ}$$

i.e. the total work done is 33.75 kJ.

8.3 Conversion of energy

As stated in section 8.1, the term ‘energy’ means ‘the capacity to do work’. Energy exists in many forms including potential, kinetic, heat, electrical, chemical, nuclear, light, and sound energy.

Potential energy is stored in a bow when the string is drawn. This is converted to kinetic energy of the arrow when it is released.

A hydro-electric pumped-storage scheme. The water is retained in the reservoir by the dam wall and is allowed to flow into the six turbines through the large pipes shown on the left-hand side. When electrical demand is low, surplus electrical energy from the grid is used to drive the turbines in reverse and pump water from the lower lake back into the reservoir.

Potential and kinetic energies are known as *mechanical energy*. Examples of potential energy include a compressed spring and water stored in the reservoir of a hydro-electric power station. A compressed spring is doing nothing, but it has the potential to do work if it is released. Likewise, while in the reservoir, the water is doing nothing; however, it has the *potential* to do work if it is released and allowed to flow downwards into the turbines.

Bodies in motion possess *kinetic energy*. For example, the water flowing downwards from the reservoir to the turbine possesses kinetic energy which can be converted into work in the turbine. Also, the rotating spindle of an electric motor possesses kinetic energy, since the motor can be used to drive devices such as machine tools or washing machines etc.

Energy can be neither created nor destroyed, but it can be converted from one form to another. The block diagram in fig. 8.6 shows examples of energy conversion.

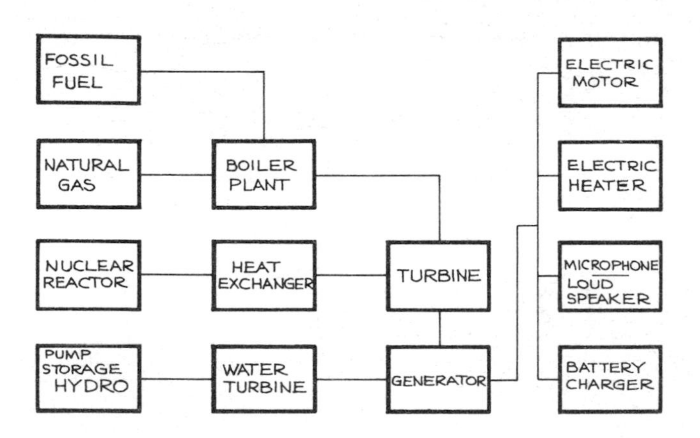

Fig. 8.6 Block diagram showing energy conversion

8.4 Efficiency and power

Referring to the block diagram in fig. 8.6, at each energy conversion (other than electrical to heat energy) the useful energy output will be smaller than the energy input. This is because some of the energy input is used to overcome friction which generates heat energy that is given up to the surroundings.

The ratio of useful energy output E_{out} to energy input E_{in} is called *efficiency*,

i.e. $$\text{efficiency} = \frac{\text{useful energy output}}{\text{energy input}} = \frac{E_{out}}{E_{in}}$$

The symbol used for efficiency is η(*eta*). Since efficiency is a ratio of like quantities, it has no units.

i.e. $\eta = \dfrac{E_{out}}{E_{in}}$

which should be remembered.

Power, P, is defined as energy, E, transferred, per unit time t. Power may also be defined as the rate at which energy is changed from one form to another.

i.e. $$\text{power} = \frac{\text{energy transfer}}{\text{time taken}}$$

or $$P = \frac{E}{t}$$

which should be remembered.

The unit for power is the watt (W). Since the unit for energy transfer is the joule (J), and the unit for time is the second (s), then

$$1\,\text{W} = 1\,\text{J/s}$$

which should be remembered.

Example 1 A load of 2 kN is raised through a height of 2.5 m in 4 seconds by an electrical power hoist. If the efficiency of the hoist is 60%, find the required input power.

It is first necessary to calculate the output power P_{out} required to raise the load, using the equation

$$P_{out} = \frac{E_{out}}{t}$$

where E_{out} is the work done on the load.

$$E_{out} = W = Fs$$

$$\therefore \quad P_{out} = \frac{Fs}{t}$$

where $F = 2\,\text{kN} = 2000\,\text{N}$ $\quad s = 2.5\,\text{m}$ and $t = 4\,\text{s}$

$$\therefore \quad P_{out} = \frac{2000\,\text{N} \times 2.5\,\text{m}}{4\,\text{s}}$$

$$= 1250\,\text{W} \quad \text{or} \quad 1.25\,\text{kW}$$

Since the hoist is not 100% efficient, the input power P_{in} is found using the relationship

$$\eta = \frac{E_{out}}{E_{in}} = \frac{P_{out}}{P_{in}}$$

$$\therefore \quad P_{in} = \frac{P_{out}}{\eta}$$

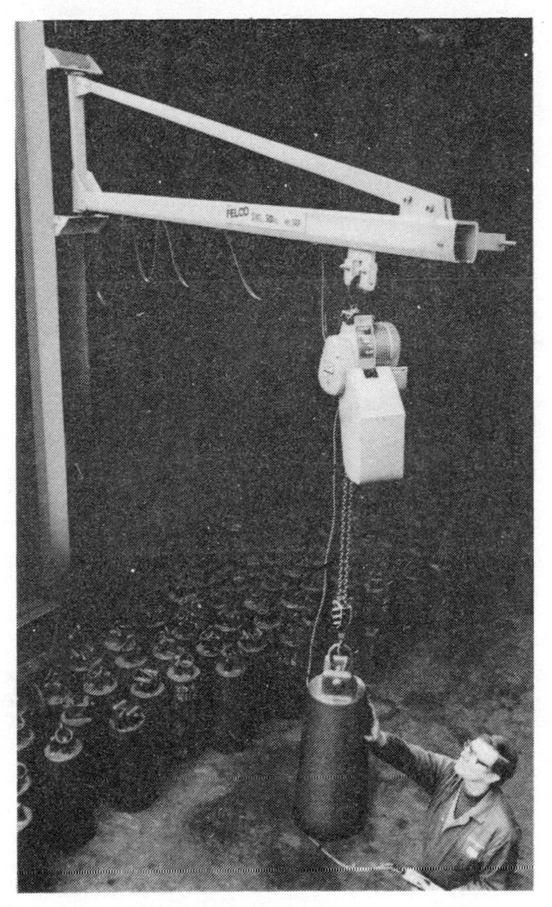

An electrically powered lifting hoist

where $\eta = 60\% = 0.6$ and $P_{out} = 1.25\,kW$

$$\therefore \quad P_{in} = \frac{1.25\,kW}{0.6}$$

$$= 2.083\,kW$$

i.e. the power input to the hoist is 2.083 kW.

Example 2 The energy in one kilogram of fuel oil is 45 MJ. The oil is supplied to a boiler which has an efficiency of 85%. In the boiler, steam is produced which is used to drive a turbine. If the efficiency of the turbine is 30% when it is producing a power output of 250 kW, calculate the mass of fuel oil required per hour in the boiler.

$$\eta = \frac{E_{out}}{E_{in}}$$

$$\therefore \quad E_{in} = \frac{E_{out}}{\eta}$$

For the turbine, $E_{out} = 250\,\text{kW}$ and $\eta = 30\% = 0.3$

$$\therefore \quad E_{in\ (turbine)} = \frac{250\,\text{kW}}{0.3}$$

$$= 833.3\,\text{kW}$$

This will be the output (E_{out}) from the boiler.

For the boiler, $\eta = 85\% = 0.85$

$$\therefore \quad E_{in\ (boiler)} = \frac{833.3\,\text{kW}}{0.85}$$

$$= 980.35\,\text{kW}$$

$$= 980.35 \times 10^3\ \text{J/s}$$

$$= 980.35 \times 10^3\ \text{J/s} \times 3600\,\text{s/h}$$

$$= 3.529 \times 10^9\ \text{J/h}$$

$$\therefore \quad \text{Mass of fuel required} = \frac{\text{energy input to boiler per hour}}{\text{energy per kilogram of fuel}}$$

$$= \frac{3.529 \times 10^9\ \text{J/h}}{45 \times 10^6\ \text{J/kg}}$$

$$= 78.42\,\text{kg/h}$$

i.e. the mass of fuel required per hour is 78.42 kg.

Exercises on chapter 8

1 Define *work*. A force of 50 kN is applied to a truck, moving it through a distance of 15 m in the direction of the force. Find the work done on the truck.

2 The work done in raising a load through a height of 6 m is 1200 J. Find the required raising force.

3 A force of 500 N is applied to a body, causing it to move in the direction of the force. If the work done on the body is 2500 J, find the distance travelled.

4 From the data given below, sketch the work diagram and use it to find the total work done.

0 to 0.5 m - force increases uniformly from 0 to 150 N;
0.5 to 1 m - force increases uniformly from 150 to 200 N;

1 to 1.5 m - force remains constant at 200 N;
1.5 to 2 m - force reduces uniformly from 200 N to zero.

5 For the work diagram shown in fig. 8.7, find the total work done.

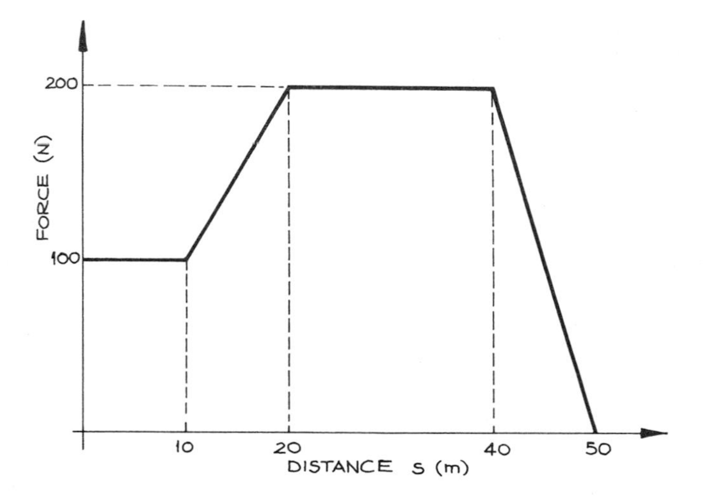

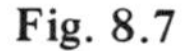
Fig. 8.7

6 A work diagram is measured and has a total area of 7200 mm^2. The force-axis scale is 1 mm ≡ 10 N, and the distance-axis scale is 1 mm ≡ 0.2 m. Find the total work done.

7 Name *five* forms of energy.

8 Energy may not be created nor destroyed but can be converted from one form to another. Fill in the missing words in each of the following statements.

a) Water in a reservoir possessess________________energy which is converted into________________energy in the turbine.

b) A rotating turbine possesses________________energy and drives an alternator set which produces________________energy.

c) A boiler uses the________________energy in the fuel to produce ________________energy in the steam which in turn drives a turbine producing________________energy.

d) A motor-car dynamo produces________________energy. When driving at night, some of this energy is used to produce________________energy.

e) The groove of a gramophone record drives the stylus, causing it to oscillate from side to side. This oscillation is converted into________________ energy which is then amplified and passed into a loudspeaker, producing ________________energy.

9 Define *efficiency*. A water turbine supplies 60 MJ of energy to a generator every minute. If the output from the generator is 950 kW, calculate its efficiency.
10 The efficiency of a lifting machine is 75%. Calculate the energy input when the energy output is 150 kJ.
11 The total energy input to a boiler is 800 GJ. If the total energy output is 500 GJ, calculate the boiler efficiency.
12 A motor-car engine produces 45 kW. Find the power available at the road-wheels when the efficiency is 80%.
13 Define *power*. A load of mass 800 kg is raised through a height of 6 m in 12 s. Calculate the power.
14 The chemical energy available in 1 kg of fuel oil is 45 MJ. The oil is used to fire a boiler having an efficiency of 80%. The boiler produces steam which drives a turbine which has an efficiency of 25%. Find the mass of fuel oil required per hour to give an output power from the turbine of 300 kW.
15 While cooking a piece of meat, the input power to a microwave cooker is 600 W. The time required to cook the meat is 38 min. In an electric oven, the input power is 3 kW and the time required to cook the same meat is 75 min. What is the efficiency of the electric oven compared with the microwave cooker?
16 In a personal stereo cassette player, the tape speed is 41.28 mm/s and the average driving force on the tape is 0.3 N. Find the total work done on the tape if it runs for 15 min. What will be the power used, assuming a drive efficiency of 80%?
17 Define *efficiency*. Explain why the output energy of a machine is less than the input energy.
18 Explain what are meant by the terms *energy* and *efficiency*.

9 Heat energy and temperature

9.1 Temperature

Temperature is a measure of hotness or coldness. For example, when the temperature is around 0°C, it *feels* cold; and when the temperature is around say 25°C, it *feels* hot.

Temperature may be measured from one of two datum points; these are absolute zero and zero on the Celsius scale. Temperature measured from absolute zero is called *absolute temperature* and is measured in units of kelvin (K). The temperature unit kelvin is one of the SI base units.

On the Celsius scale, zero corresponds with 273.15 K (usually taken to be 273 K) on the absolute temperature scale and is the temperature at which ice starts to melt at atmospheric pressure. Temperature on the Celsius scale is measured in degrees Celsius (°C). The term 'degrees centigrade' is also used.

It is important to note that a temperature interval of 1°C is the same as a temperature interval of 1 K,

i.e. a change in temperature of 1°C = a change in temperature of 1 K

9.2 State of a substance

A substance can exist in three *states* or *phases* – i.e. solid, liquid, or gaseous (vapour). The state of a substance depends on its temperature and pressure. For example, water at atmospheric pressure is solid (ice) at temperatures below 0°C, liquid between 0°C and 100°C, and vapour (steam) above 100°C.

9.3 Sensible and latent heat

Heat is a form of energy. When heat is supplied to a substance, the temperature of the substance may rise or the substance may change its state. Similarly, if heat is *rejected* by a substance, the temperature of the substance may *fall*, or, again the substance may change its state.

Heat energy which causes the temperature of a substance to rise or fall is called *sensible heat*, the word 'sensible' meaning 'perceptible or evident to the senses' – i.e. the temperature rise or fall can be *seen* on a thermometer.

Heat energy which causes a change of state to occur is called *latent heat*, the word 'latent' meaning 'hidden' or 'not shown' – i.e., during a change of state, the thermometer registers no change in temperature.

9.4 Specific heat capacity

Specific heat capacity is defined as the amount of sensible heat energy required to give unit mass of a substance a temperature rise of one degree. For example, one kilogram of water requires 4187 joules of heat energy to raise its temperature by one degree Celsius. Thus the specific heat capacity of water is 4187 joules per kilogram per degree Celsius. This value is normally stated as 4187 J/(kg °C).

The symbol used for specific heat capacity is c.

Table 9.1 gives typical values of the specific heat capacity, c, for various substances.

Table 9.1 Typical values of specific heat capacity

Substance	Specific heat capacity (J/(kg °C))
Water	4187
Aluminium	921
Cast iron	544
Steel	494
Wrought iron	473
Copper	385
Mercury	138
Lead	130

9.5 Quantity of heat energy

The amount or quantity of heat energy, Q_s, to produce an increase in temperature (i.e. sensible heat) of a substance depends upon

a) the mass, m, of the substance – the larger the mass, the greater the heat energy required;
b) the specific heat capacity, c, of the substance;
c) the temperature change, $\Delta\theta$ – the greater the change in temperature, the larger the quantity of heat energy required.

Thus, $\text{quantity of heat energy} = \text{mass} \times \text{specific heat capacity} \times \text{temperature change}$

or $$Q_s = mc\Delta\theta$$

which should be remembered.

The unit for quantity of heat energy is the joule (J). The units kilojoule ($1\,\text{kJ} = 1 \times 10^3\,\text{J}$) and megajoule ($1\,\text{MJ} = 1 \times 10^6\,\text{J}$) are also used

Example 1 A copper ingot of mass 6 kg is heated in a furnace, increasing its temperature from 16°C to 610°C. Find the quantity of sensible heat supplied. Take the specific heat capacity of copper as 385 J/(kg °C).

$$Q_s = mc\Delta\theta$$

where $m = 6\,\text{kg}$ $\quad c = 385\,\text{J/(kg °C)}$ and $\Delta\theta = 610°\text{C} - 16°\text{C} = 594°\text{C}$

$$\therefore \quad Q_s = 6\,\text{kg} \times 385\,\text{J/(kg °C)} \times 594°\text{C}$$

$$= 1.372 \times 10^6\,\text{J} \quad \text{or} \quad 1.372\,\text{MJ}$$

i.e. the sensible heat supplied is 1.372 MJ.

Example 2 Four litres of water at a temperature of 16 °C are supplied with 200 kJ of heat energy. If the mass of one litre of water is 1 kg and its specific heat capacity is 4.2 kJ/(kg°C), calculate the final temperature of the water.

$$Q_s = mc\Delta\theta$$

$$\therefore \quad \Delta\theta = \frac{Q_s}{mc}$$

where $Q_s = 200\,\text{kJ} = 200 \times 10^3\,\text{J}$ $\quad m = 4 \text{ litres} \times 1\,\text{kg/litre} = 4\,\text{kg}$

and $c = 4.2\,\text{kJ/(kg°C)} = 4200\,\text{J/(kg°C)}$

$$\therefore \quad \Delta\theta = \frac{200 \times 10^3\,\text{J}}{4\,\text{kg} \times 4200\,\text{J/(kg°C)}}$$

$$= 11.9\,°\text{C}$$

But $\Delta\theta$ = final temperature – initial temperature

$\therefore$ final temperature = $\Delta\theta$ + initial temperature

where initial temperature = 16°C

$\therefore$ final temperature = 11.9°C + 16°C

= 27.9°C

i.e. the final temperature of the water is 27.9°C.

Example 3 A steel component at a temperature of 760 °C is immersed in a tank containing 60 kg of water where its temperature is reduced to 20°C. Initially, the water in the tank was at a temperature of 16°C. Assuming that the final temperature of the water is 20 °C and that all the heat lost by the component is absorbed by the water, find the mass of the steel component. For water, $c_w = 4200\,\text{J/(kg °C)}$; for steel, $c_s = 500\,\text{J/(kg °C)}$.

Heat energy gained by water, $Q_w = m_w c_w \Delta\theta_w$

where $m_w = 60\,\text{kg}$ $\quad c_w = 4200\,\text{J/(kg °C)}$

and $\Delta\theta_w = 20\,°\text{C} - 16\,°\text{C} = 4\,°\text{C}$

$$\therefore \quad Q_w = 60\,\text{kg} \times 4200\,\text{J/(kg°C)} \times 4\,°\text{C}$$

$$= 1.008 \times 10^6\,\text{J}$$

Heat energy lost by steel, $Q_s = m_s c_s \Delta\theta_s$

$$\therefore \quad m_s = \frac{Q_s}{c_s \Delta\theta_s}$$

where $Q_s = Q_w = 1.008 \times 10^6\,\text{J}$ $\quad c_s = 500\,\text{J/(kg °C)}$

and $\Delta\theta_s = 760\,°\text{C} - 20\,°\text{C} = 740\,°\text{C}$

$$\therefore \quad m_s = \frac{1.008 \times 10^6 \text{ J}}{500\text{ J/(kg°C)} \times 740\text{°C}}$$

$$= 2.72\text{ kg}$$

i.e. the mass of the component is 2.72 kg.

9.6 Specific latent heat

The specific latent heat of a substance is defined as the amount of heat energy required to cause a mass of one kilogram of the substance to change from one state to another without a change in temperature.

When the change of state is from a liquid to a vapour, the term *specific latent heat of vaporisation* is used, while the term *specific latent heat of fusion* is used when the change of state is from solid to liquid. For example, at atmospheric pressure, one kilogram of water at a temperature of 100°C requires 2257 kJ of heat energy to convert it into steam at the same temperature - i.e. the specific latent heat of vaporisation of water is 2257 kJ/kg.

The symbol used for specific latent heat is h.

Table 9.2 gives typical values of the specific latent heat h for various substances at atmospheric pressure.

Table 9.2 Typical values of specific latent heat at atmospheric pressure

	Specific latent heat (kJ/kg)	
Substance	Fusion	Vaporisation
Water	334	2257
Ammonia	—	1369
Mercury	—	290
Carbon dioxide (CO_2)	—	364
Oxygen	—	243
Aluminium	9130	
Copper	3850	
Lead	1260	
Steel	4200	

It should be noted that, at the same temperature, changes of state will occur in the opposite direction - i.e. vapour to liquid or liquid to solid - when an amount of heat energy equal to the latent heat is extracted from one kilogram of the substance.

9.7 Quantity of heat to produce a change of state

When heat energy is absorbed by a substance and there is no rise in temperature or pressure, then the substance must be changing its state.

The quantity of heat energy, Q_l (i.e. latent heat), required to change the state of the substance depends upon

a) the mass, m, of the substance - the larger the mass, the greater the latent heat energy required;
b) the specific latent heat, h, of the substance.

Thus, quantity of latent heat energy = mass $\times$ specific latent heat

or $$Q_l = mh$$

which should be remembered.

Example 1 The specific latent heat required to melt ice is 335 kJ/kg. Find the quantity of latent heat required to melt 3 kg of ice at 0 °C.

$$Q_l = mh$$

where $m = 3\,\text{kg}$ and $h = 335\,\text{kJ/kg} = 335 \times 10^3\,\text{J/kg}$

$$\therefore \quad Q_l = 3\,\text{kg} \times 335 \times 10^3\,\text{J/kg}$$
$$= 1.005 \times 10^6\,\text{J} \quad \text{or} \quad 1.005\,\text{MJ}$$

i.e. the quantity of latent heat required is 1.005 MJ.

Example 2 Eight kilograms of water at a temperature of 50 °C are heated until they are converted into steam at a temperature of 120 °C, the change in state occurring at a temperature of 100 °C. If the specific heat capacity of the water and the steam are 4187 J/(kg °C) and 2742 J/(kg °C) respectively, and the specific latent heat is 2257 kJ/kg, calculate the total quantity of heat energy required.

$$Q_{\text{total}} = Q_s \text{ (for the water)} + Q_s \text{ (for the steam)} + Q_l$$

Let suffix w refer to the water and suffix g refer to the steam, then

$$Q_{\text{total}} = mc_w\Delta\theta_w + mc_g\Delta\theta_g + mh$$
$$= m(c_w\Delta\theta_w + c_g\Delta\theta_g + h)$$

where $m = 8\,\text{kg}$ $\quad c_w = 4187\,\text{J/(kg °C)}$

$\Delta\theta_w = (100\,°\text{C} - 50\,°\text{C}) = 50\,°\text{C}$ $\quad c_g = 2742\,\text{J/(kg °C)}$

$\Delta\theta_g = (120\,°\text{C} - 100\,°\text{C}) = 20\,°\text{C}$

and $h = 2257\,\text{kJ/kg} = 2257 \times 10^3\,\text{J/kg}$

$$\therefore \quad Q_{\text{total}} = 8\,\text{kg} \times [4187\,\text{J/(kg°C)} \times 50\,°\text{C} + 2742\,\text{J/(kg°C)} \times 20\,°\text{C} + 2257 \times 10^3\,\text{J/kg}]$$
$$= 20.17 \times 10^6\,\text{J} \quad \text{or} \quad 20.17\,\text{MJ}$$

i.e. the total quantity of heat energy required is 20.17 MJ.

9.8 Cooling curves

If, during the heating or cooling of a substance, a plot of temperature against time is made, the resulting graph is known as a *heating* or *cooling curve* for the substance. Cooling curves are particularly useful since they can be used to find the temperature at which a substance changes state - usually from a liquid to a solid. The following experiment can be done at home in the kitchen.

Experiment 9.1 To plot the cooling curve for candle wax.

Place a cup containing solid candle wax in a saucepan of water and bring the water to the boil. Adjust the heat under the saucepan so that the water is boiling gently, i.e. it is simmering. Heating should continue until all the wax has melted, when - taking great care - the cup can be removed from the water.

Suspend a mercury-in-glass thermometer with a scale range of 0 to 100 °C over the cup so that the bulb is immersed in the liquid wax as shown in fig. 9.1. Note the temperature at time $t = 0$. Now take temperature readings at one-minute intervals until the wax has completely solidified.

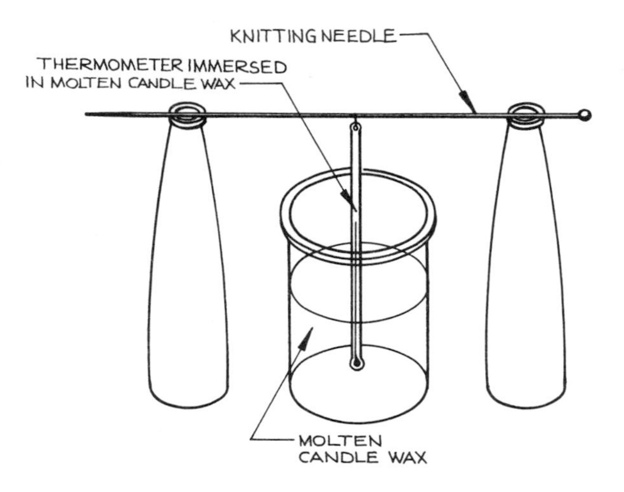

Fig. 9.1 Experiment on candle wax

The results from such an experiment are shown in Table 9.1, from which the graph in fig. 9.2 has been drawn.

Table 9.1 Temperature–time results for cooling candle wax

Time (min)	0	1	2	3	4	5	6	7	8	9	10	11	15
Temp. (°C)	62	61	60	59	58	57	56	55.8	55.4	55.3	55.2	55	55
Time (min)	16	17	18	19	20	21	22	23	24	25	30	35	
Temp. (°C)	54.8	54.7	54.5	53.5	52.5	52	51.5	51	50.6	50.5	49.5	48.5	

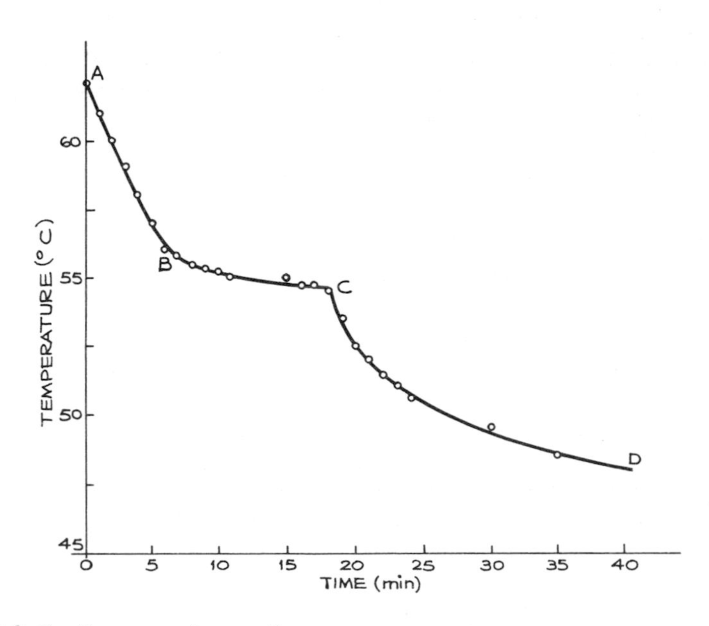

Fig. 9.2 Cooling curve for candle wax

Referring to the graph in fig. 9.2, between A and B the liquid wax is cooling at a uniform rate and sensible heat energy is being transferred to the surroundings. Between B and C the wax is solidifying at nearly constant temperature and latent heat energy is being transferred to the surroundings. Between C and D, the solid wax is cooling more slowly and the heat energy transferred to the surroundings is sensible heat. As a conclusion to this experiment, we can say that the solidification or 'freezing' temperature of the candle wax used is approximately 55 °C.

To avoid scalding, great care must be taken when removing the cup from the boiling water during this experiment.

Similar experiments can be carried out in the physical-science laboratory using pure metals such as tin or lead. The shape of the cooling curves for these substances will be similar to the one shown in fig. 9.2.

Cooling curves for alloys - i.e. metals or substances containing two or more ingredients - are usually more complex. Figure 9.3 shows a cooling curve for solder, which is an alloy of lead and tin. Because lead and tin solidify at different temperatures, there will be a fall in temperature with respect to time during solidification, as is shown between B and C in fig. 9.3.

9.9 Effect of temperature change on solids and liquids

When any substance - solid or liquid - absorbs sensible heat energy, its temperature will rise and it will *expand* (grow larger) in *all* directions. It

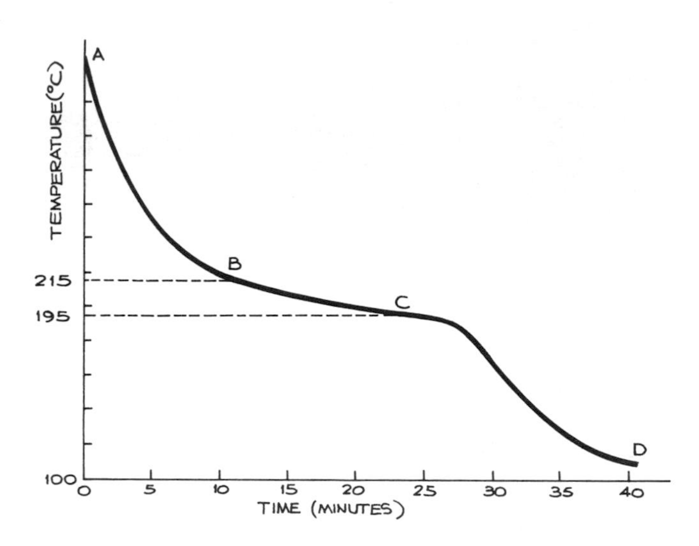

Fig. 9.3 Cooling curve for solder

follows that, when sensible heat energy is transferred from the substance, the opposite will occur - i.e. the temperature will fall and the substance will contract (grow smaller) in *all* directions.

One exception to this is water between 0 and 4°C. Between these temperatures, the water *contracts* with *increase* in temperature and *expands* with *decrease* in temperature. Unfortunate side effects of this phenomenon are the problems which occur when water freezes in a pipeline or an engine cylinder block. The expansion causes cracking to occur which is discovered only when the ice has melted!

The fact that materials expand and contract with changes in temperature is used to advantage in engineering manufacture. For example, to fit the toothed starter ring on to the rim of the motor-car engine flywheel shown in fig. 9.4, the flywheel is cooled to a very low temperature, causing it to contract, and the ring is then pressed into place. When the temperature of the flywheel increases, it expands - thus gripping the ring very tightly.

Expansion and contraction of materials is also put to good use in such devices as thermostats and liquid-in-glass thermometers. Figure 9.5 shows a thermostatic control valve for a domestic central-heating radiator. In this device, the position of the water valve is controlled by the rod B. As the temperature in the room increases, the rod expands until the water valve closes, cutting off the water supply to the radiator. When the room temperature falls, the rod contracts and the water valve reopens. The temperature at which the water valve closes may be increased or decreased by varying the initial gap between the end of the rod and the valve plunger - i.e., to increase the temperature, the initial gap is made larger so that the rod has to expand a greater distance to close the valve.

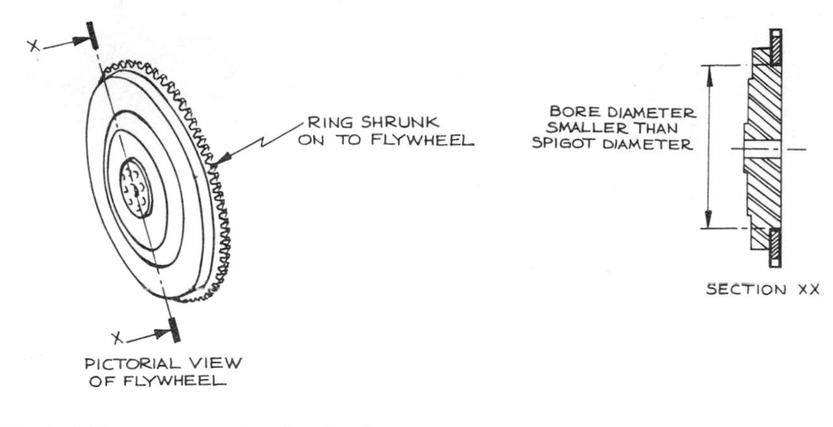

Fig. 9.4 Motor-car engine flywheel

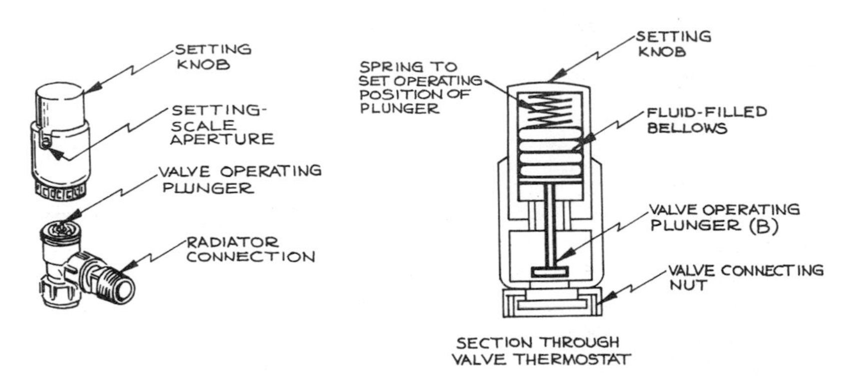

Fig. 9.5 Thermostatic control valve for a radiator

In the liquid-in-glass thermometer shown in fig. 9.6, the thin glass bulb contains liquid (mercury in this case) which, with temperature change, is free to expand or contract along the fine-bore glass tube, called a capillary tube.

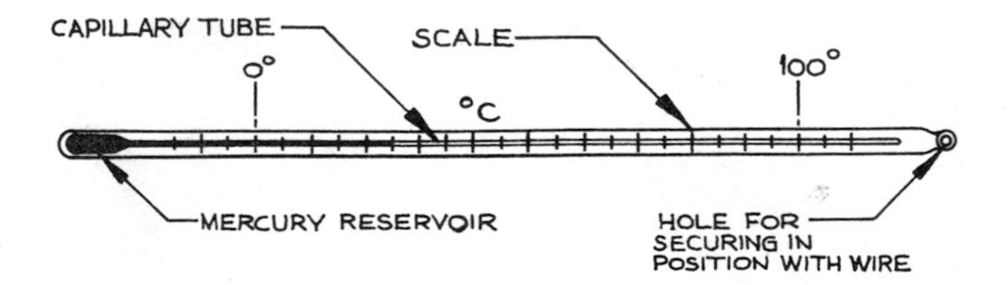

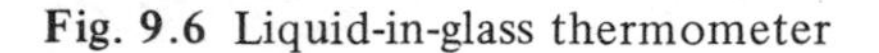
Fig. 9.6 Liquid-in-glass thermometer

An expansion joint in a multi-storey car park

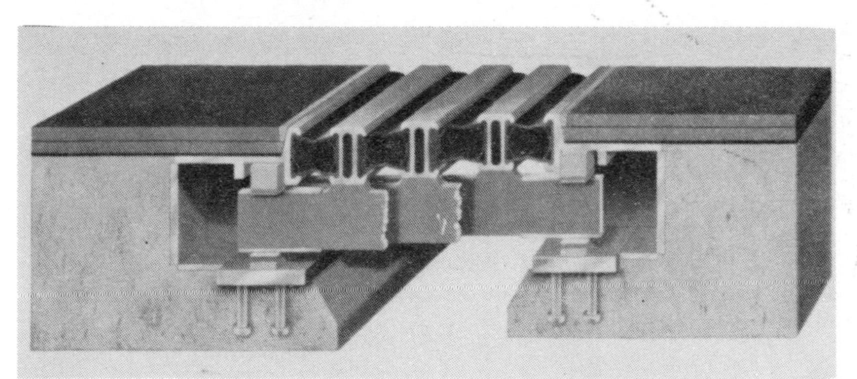
A cross-section of the expansion joint shown opposite. Note the expansion gap at each end of the centre section

Expansion and contraction of materials also have many disadvantages. For example, expansion must be allowed for in road bridges and motorway flyovers, steel-framed buildings, and railway lines. Figure 9.7 shows a method used to allow for expansion and contraction in steam pipelines.

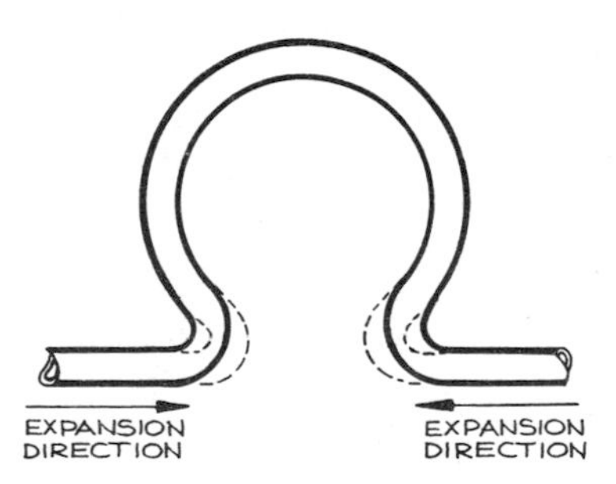

Fig. 9.7 Expansion loop in a steam pipeline

Exercises on chapter 9

1 Define *specific heat capacity*. The temperature of 2 kg of copper is increased from 15 °C to 125 °C. Find the heat energy supplied. The specific heat capacity of copper is 380 J/(kg °C).

2 A steel billet at a temperature of 720 °C is cooled rapidly in water until its temperature is 20 °C. During cooling, the billet gives up 520 kJ of heat energy. If the specific heat capacity of the steel is 494 J/(kg °C), find the mass of the billet.

3 A pure metal in the molten state is allowed to cool and solidify. Sketch the cooling curve, indicating on it where sensible or latent heat energy is being rejected.

4 The specific heat capacity of water is 4187 J/(kg °C). Calculate the amount of heat energy required to raise the temperature of 25 kg of water by 60 °C.

5 State the difference between *sensible* heat and *latent* heat.

6 Explain what are meant by the terms *temperature*, *heat*, and *state* of a substance.

7 A copper beaker has a mass of 60 g and contains 120 g of water. Submerged in the water is a steel ball-bearing of mass 30 g. Find the total heat energy required to increase the temperature of the beaker, water, and ball-bearing from 12 °C to 48 °C. Assume no heat loss to the surrounding air. The specific heat capacities are 4190 J/(kg °C) for water, 380 J/(kg °C) for copper, and 490 J/(kg °C) for steel.

8 Find the heat energy required to convert 5 kg of ice at a temperature of –5 °C into water at a temperature of 10 °C. The specific heat capacities of water and of ice are 4190 J/(kg °C) and 2190 J/(kg °C) respectively, and the latent heat of fusion is 330 kJ/kg.

9 Six kilograms of water at a temperature of 80 °C is converted into superheated steam at a temperature of 130 °C, the change of state taking place at 100 °C. Find the heat energy required, given that the specific heat capacities of water and of superheated steam are 4.187 kJ/(kg °C) and 2.02 kJ/(kg °C) respectively, and the latent heat of vaporisation is 2257 kJ/kg.

10 Sketch the form of the cooling curve which would be obtained when converting the following from a liquid into a solid: (a) a pure metal, (b) an alloy of two metals. Explain why the curves are not the same shape.

11 A pure molten substance at a temperature of 400 °C was allowed to cool slowly until its temperature was 20 °C. During cooling, it was observed that freezing occured when the temperature was 320 °C. Sketch the cooling curve and calculate the total heat energy lost to the surroundings, given that the mass of the substance was 60 g, the specific heat capacity of the molten substance was 480 J/(kg °C) and that of the solid substance was 240 J/(kg °C), and the latent heat of fusion was 65 J/kg.

12 Copper rivets having a total mass of 1.6 kg and at a temperature of 20 °C are heated in a furnace until their temperature is 560 °C. Find the heat energy required if the specific heat capacity of copper is 390 J/(kg °C). If, during heating, the power used is 351 W, calculate the time in minutes that the rivets are in the furnace, assuming that 80% of the available heat energy is absorbed by the rivets.

10 Electricity

10.1 Introduction

In the study of electricity, we are concerned with electric current as a basic quantity.

An electric current is a flow of electrons through a conductor such as a copper wire.

In any material there are some electrons which have broken free from the parent atom. These free electrons possess a negative charge and are therefore able to act as charge carriers.

There are many more free electrons in a materials such as copper than there are in glass, say. Copper is thus referred to as a good *conductor* while glass is referred to as a good *insulator*.

When an electric cell, such as a torch battery, is connected across the opposite ends of a conductor, the negatively charged free electrons are attracted by the positive terminal of the cell and so move towards it along the wire. Electrons are in turn fed into the wire at the negative terminal of the cell.

An insulator is simply a very bad conductor of electricity, since hardly any electrons can flow through it.

The letter I is used to represent electric current, and current is represented in electric circuits as shown in fig. 10.1.

I

Fig. 10.1 Representation of electric current in circuits

Electric current is a measure of the rate of flow of electrons through a conductor. The unit is the ampere (symbol A). The ampere is an SI base unit, which means that other units are defined in terms of this unit.

The effect of the electric cell causing current to flow through a conductor may be likened to the way in which a mechanical pump causes water to flow through a pipe. The cell is said to produce an *electromotive force* (e.m.f.) which causes current to flow.

The letter E is used to represent e.m.f., and e.m.f. is represented in electric circuits as shown in fig. 10.2. The unit of e.m.f. is the volt (symbol V).

An electric cell, such as the dry cell used in a torch, or the lead-acid cell used in a motor car, produces an e.m.f. by chemical means (as discussed in

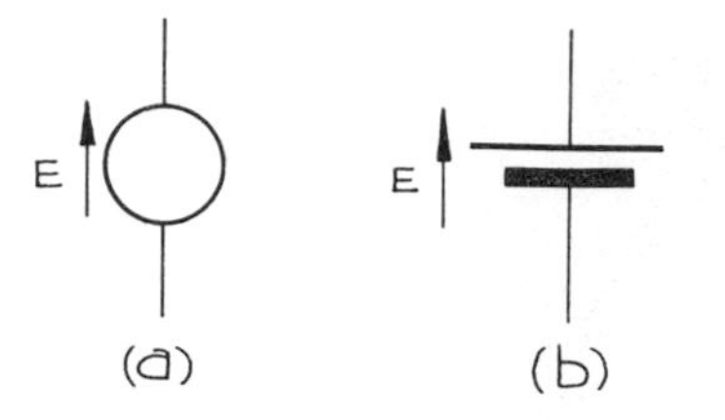

Fig. 10.2 Representation of electromotive force (e.m.f.) in circuits – (a) and (b) are alternatives.

chapter 12), but there are other ways of producing an e.m.f. Most of the electricity used in industry and in the home is produced in power stations using electric generators (or alternators) which make use of the electromagnetic effect of an electric current, which we will discuss briefly in section 13.4.

The electricity supplied to homes and industries is referred to as a.c. (alternating current), since the voltage alternates in direction, going both above and below the zero at a frequency of 50 Hz with a waveform like that in fig. 7.1.

The electricity supplied by an electric cell is referred to as d.c. (direct current), since the voltage is always unidirectional, i.e. in one direction only.

10.2 Electricity today

Nowadays we are familiar with electricity as something that we can make use of simply by plugging the chosen electrical appliance into a wall socket. Electricity is used in a wide range of industrial, domestic, military, and social situations, and it is useful to consider a few of them.

The manufacturing industry uses electric motors for driving lathes, planes, and milling and drilling machines as well as in applications such as weaving, food-mixing, steel-rolling, and paper-making.

Electric drives are used in electric trains, some overhead railcars, and cable cars and will in the near future be used in electric motor cars for medium-distance travel.

Electric heating, as well as being used for keeping us warm, is used in electric furnaces in foundries etc.

In the communications industry, electricity is used to power transmitters and receivers. 'Communications' is the general term used to include the broadcasting industry of radio and television as well as telephony, telemetry (the transmission of instrument readings to a remote receiver), and satellite communications.

Computers work by using electric signals which are handled at more than a million a second. This makes them ideal for performing calculations, processing data such as in banks and wages offices, and controlling industrial processes.

The Sinclair C5 battery-powered vehicle, using a 250 W electric motor (as used in a modern washing machine) and powered by a 36 A h lead–acid battery

Robots used in the production of the Sierra at Ford's Dagenham plant

A 12V d.c. starter motor as used in a modern motor car, incorporating the starter solenoid

Many process-control systems such as those used in the chemical manufacturing industry use electrical methods of sensing and controlling fluid flow, temperature, pressure, etc.

Hospitals use electricity for heating and lighting, and for intensive-care monitoring equipment, electrocardiac equipment, and body-scan, brain scan, and X-ray machines.

In the home, electricity is used for lighting, cooking, radios, TV's, central heating and some water heating, telephones, electric bells, toasters, refrigerators, washing machines, electric mixers, etc.

In the military field, electricity is used in aircraft, ships, and land vehicles – for instrumentation, radar, computing, control, and communications.

Construction sites use temporary electrical installations for lighting, hoists, and electric cement mixers.

Electricity is used in the motor-car industry for powering and controlling robot assembly lines, electric arc welding, cranes, and electroplating.

In all, it is clear that there is a tremendously wide area of usage of electricity.

10.3 Conductors and insulators

We have said that electric current is a flow of electrons along a wire, rather like water flow through a pipe.

Two 400 kV cable samples: a 400 kV Thames river crossing overhead cable (left) and a 400 kV underground cable (right)

Conductors are materials which present little resistance to this flow of electric current. Examples of good conductors are silver, copper, and aluminium. Silver is too expensive for general use, and cables are thus made from copper or aluminium. They vary in size, from the fine multi-strand wire used to interconnect electronic components, to the heavy overhead and underground cables used to transmit and distribute electrical power.

Other materials such as steel and iron will act as conductors, but they do not conduct as easily as do copper and aluminium.

Some materials are very poor conductors of electricity and they are referred to as insulators. Good insulators are air, paper, plastic, ceramic (pot), mica, and glass.

Insulators are used to prevent the flow of electric current. Ceramic insulators are used to insulate high-voltage overhead cables from their support towers. P.V.C. and p.t.f.e. are used as an insulating cover for electrical cables, and paper is used to insulate windings in transformers.

10.4 Effects of electric current

When a current flows through a conductor, several effects occur.

Firstly there is the *heating effect*. When a current flows through a conductor, the conductor heats up. This is, of course, the effect used in say the domestic immersion heater or electric fire.

Secondly there is the *electrochemical effect*. This occurs when the electric current is made to flow through a liquid electrolyte (a conducting solution), such as copper-sulphate solution, by connecting an e.m.f. across plates dipping into the electrolyte. The result is that copper is deposited on one of the plates and removed from the other plate. This effect is made use of in the electroplating industry and will be discussed in more detail in chapter 12.

Thirdly there is the *electromagnetic effect*. A current flowing through a conductor will produce a magnetic field around the conductor. This effect is used in motors, generators, transformers, etc. and is discussed in chapter 13.

10.5 Potential difference

Potential difference (p.d.) in an electric circuit may be likened to pressure difference in a fluid circuit. For fluid to flow through a pipe, there must be a difference in pressure between the two ends of the pipe.

Similarly, for electric current to flow between two points in a conductor, there must be a potential difference between those two points. When a p.d. exists between any two points in an electric circuit, then current will flow.

Potential difference has the same units as e.m.f., i.e. volts.

Potential difference may be measured between any two points in the circuit, while e.m.f. refers to a source of electric current, such as an electric cell or an electric generator.

10.6 Electrical resistance

In electric circuits, the term *resistance* means resistance to flow of electric current. A high resistance will allow only a small current to flow, while a low resistance will allow a large current to flow. This may be likened to the way in which the bore and length of a pipe affect the flow rate of fluid through the pipe.

All electric circuits have some resistance. The resistances of electric-cooker elements and lamp filaments are low - all heating and lighting elements have a low resistance.

In electric circuits, resistive components are made with a wide range of values to limit the flow of electric current. They are referred to as 'resistors'. Some resistors are made of wire wound around a former, while others are made from carbon or metal-oxide compounds formed into a solid rod.

10.7 The unit of resistance

The letter R is used as the symbol for resistance, and resistance is represented in electric circuits as shown in fig. 10.3. For a given resistance, the current that flows through the resistor is in proportion to the potential difference across it.

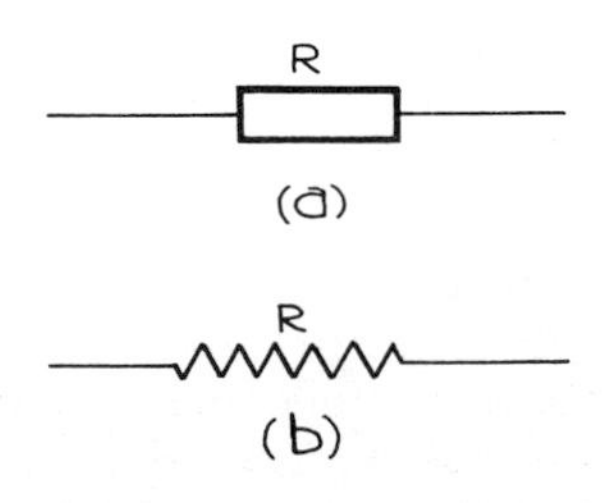

Fig. 10.3 Representation of resistance in circuits - (a) and (b) are alternatives

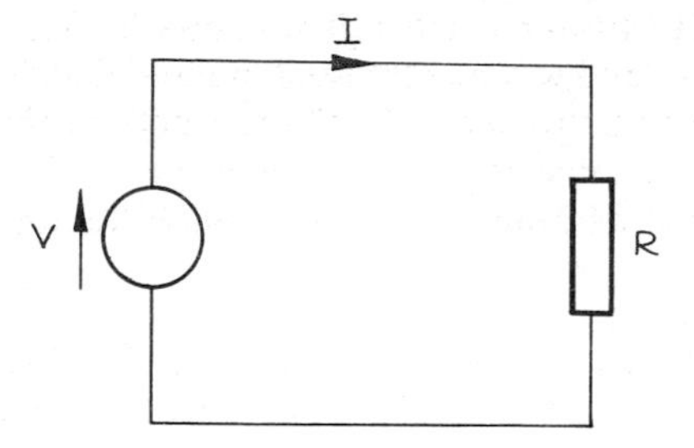

Fig. 10.4 A circuit in which a potential difference V is connected across a resistance R

Consider the circuit shown in fig. 10.4 in which a potential difference. V is connected across a resistance R. As the p.d. is increased from zero, the current I will increase proportionally (i.e. if V is doubled then I is also doubled).

Plotting a graph of V against I gives the characteristic (i.e. the relationship between the two quantities) shown in fig. 10.5 (a). Notice that the graph is a straight line. The slope of the graph is termed the resistance of the conductor.

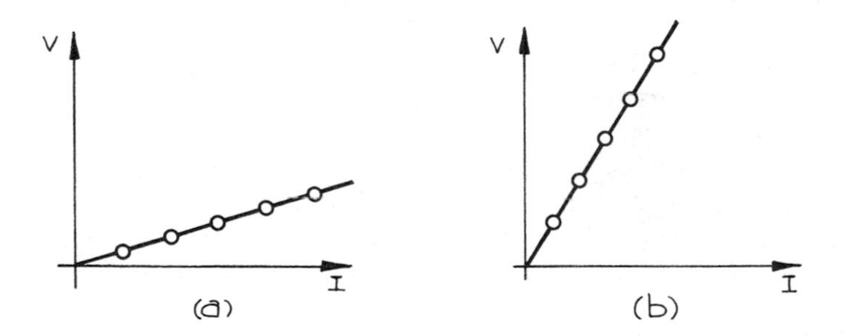

Fig. 10.5 Graphs of V against I for the circuit of fig. 10.4 with (a) a small resistor, (b) a large resistor

Resistance is thus the ratio of the potential difference (V) across the conductor to the current (I) flowing through it:

$$R = \frac{V}{I}$$

Figure 10.5 (b) shows the graph for a different resistor. The slope of graph (b) is greater than that of (a), and thus the resistance used to obtain graph (b) is greater than that of (a).

In the equation

$$R = \frac{V}{I}$$

when $V = 1$ volt and $I = 1$ ampere then R is defined as 1ohm (symbol Ω).

i.e. $1 \text{ ohm} = \dfrac{1 \text{ volt}}{1 \text{ ampere}}$

or $1\,\Omega = 1\,\text{V/A}$

i.e. the ohm is defined as the resistance between two points such that a potential difference of one volt between the two points causes a current of one ampere to flow.

The equation for resistance is often referred to as Ohm's law and written in the form

$$V = IR$$

i.e. if a resistance R ohms carries a current I amperes, then the potential difference V volts across the resistor is given by the product of I multiplied by R.

Example A heating element of an electric cooker has a resistance of 60 Ω. Calculate the current taken if the element is connected to a mains supply of 240 V.

$$I = \frac{V}{R}$$

where $V = 240\,\text{V}$ and $R = 60\,\Omega$

$$\therefore \quad I = \frac{240\,\text{V}}{60\,\Omega}$$

$$= 4\,\text{A}$$

i.e. the cooker element takes a current of 4 A.

10.8 Non-linear resistance

In some components the variation of current with voltage does not produce a linear graph. These components are referred to as non-linear. One such component is a lamp.

Figure 10.6 shows a potential difference connected across a lamp. When the p.d. is increased from zero, the current increases, and this current causes the lamp filament to heat up. As we shall see in section 10.17, the resistance of a metal increases with temperature; so, as the lamp filament (which is made from the metal tungsten) heats up, its resistance will increase.

Thus, as the p.d. across the lamp is increased, the current increases – but not linearly, due to the changing resistance of the lamp with current. The

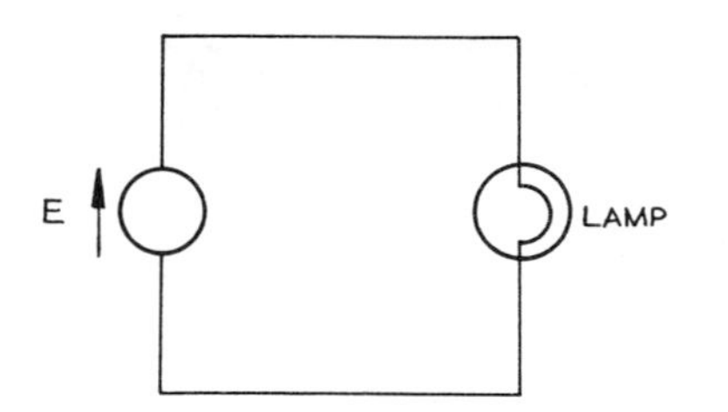

Fig. 10.6 A potential difference connected across a lamp

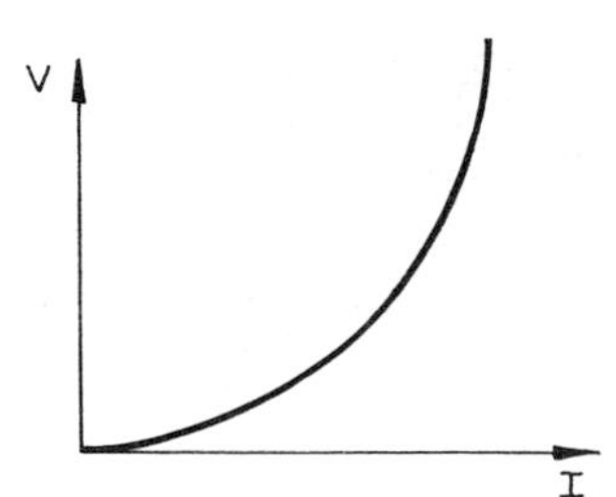

Fig. 10.7 Graph of p.d. against current for the circuit of fig. 10.6

higher the current, the higher the lamp resistance. The resulting graph of p.d. against current is shown in fig. 10.7.

10.9 Standard symbols for electrical components

A range of standard symbols is used to represent electrical components and instruments when drawing circuit diagrams. These are given in British Standard 3939.

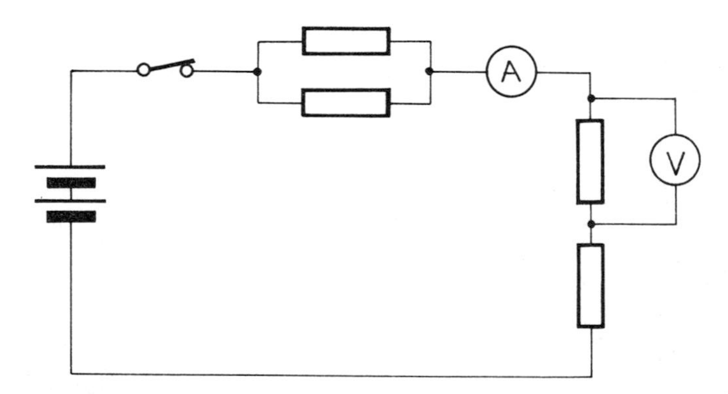

Fig. 10.8 A circuit diagram using British Standard graphical symbols

In the circuit of fig. 10.8, for example, the diagram consists of symbols. It is important to use the correct symbols when drawing a circuit diagram, and to be able to interpret the diagram when wiring an industrial installation or building or fault-finding an electric circuit.

Some of the more commonly used symbols are shown in fig. 10.9.

10.10 Ammeters and voltmeters

Electric current is measured with an instrument called an *ammeter*. E.m.f. and potential difference are measured with a *voltmeter*.

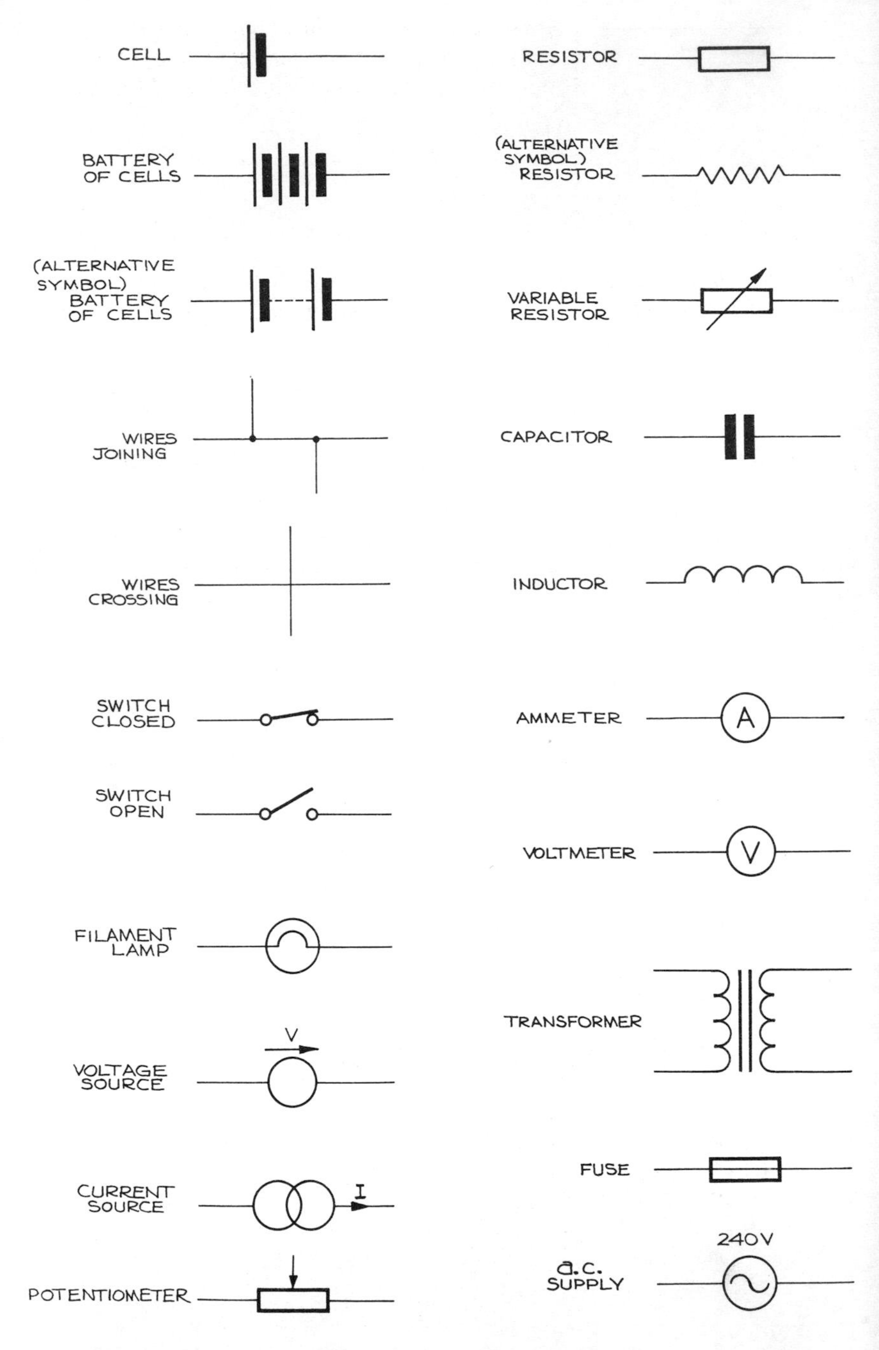

Fig. 10.9 A range of the more commonly used British Standard graphical symbols

Current flows *through* a circuit, and therefore the ammeter must be connected in the circuit as shown in fig. 10.10. The current flowing through the circuit will then flow through the ammeter. The ammeter is said to be 'in series' with the circuit.

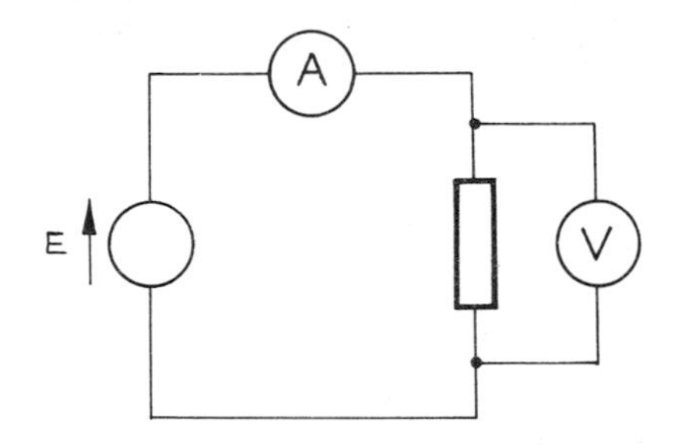

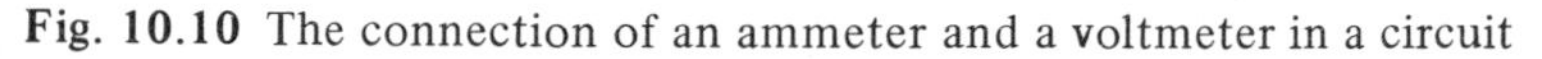

Fig. 10.10 The connection of an ammeter and a voltmeter in a circuit

A voltmeter measures e.m.f. and p.d. The p.d. across a resistor may be measured by connecting the voltmeter across the resistor as shown in fig. 10.10. The voltmeter is said to be connected 'in parallel' with the resistor.

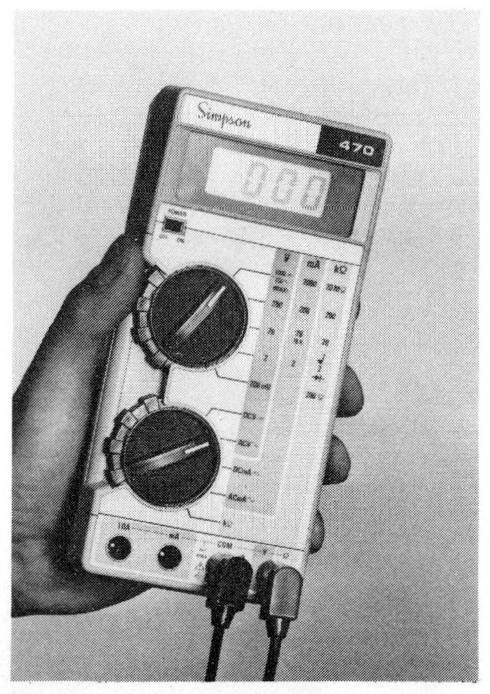

An electronic hand-held multimeter with switch selections allowing measurement of current, voltage, and resistance

An ammeter must have a low resistance so that it has negligible effect on the current being measured. It should therefore never be connected directly across an electric source or else a large current would flow and the instrument would be damaged.

A voltmeter, on the other hand, must have a very high resistance, so that it takes negligible current and so does not affect the potential difference being measured.

10.11 Electric circuits

A simple electric circuit is shown in fig. 10.11, where a cell of e.m.f. E is connected across a resistance R. The current I which flows is, by Ohm's law,

$$I = \frac{E}{R}$$

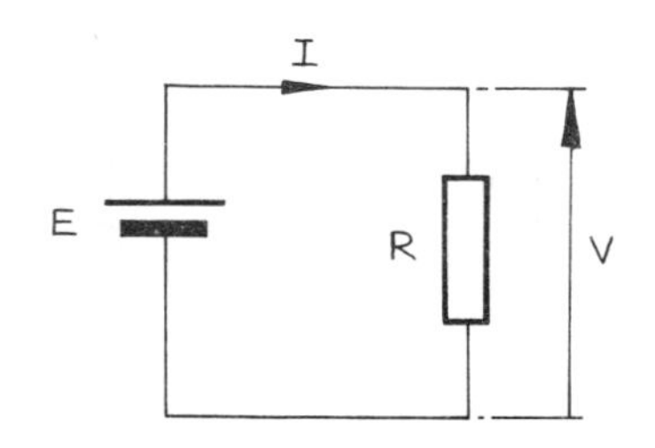

Fig. 10.11 A simple electric circuit

For current to flow, the circuit must consist of a complete closed path, otherwise it is referred to as an 'open circuit' and no current will flow.

We say that a potential difference exists across the resistance R, given by

$$V = IR$$

In the case shown in fig. 10.11, the potential difference V across the resistance is equal to the source e.m.f. E.

Experiment 10.1 To investigate the relationship between current and (a) a resistor, (b) a non-linear component such as a lamp.

The circuit is as shown in fig. 10.12.

Equipment used

a) A variable 0 to 12 V d.c. supply capable of supplying 2 A
b) A d.c. ammeter
c) A d.c. voltmeter
d) A 24 W wire-wound resistor
e) A 12 V 24 W car side-light bulb

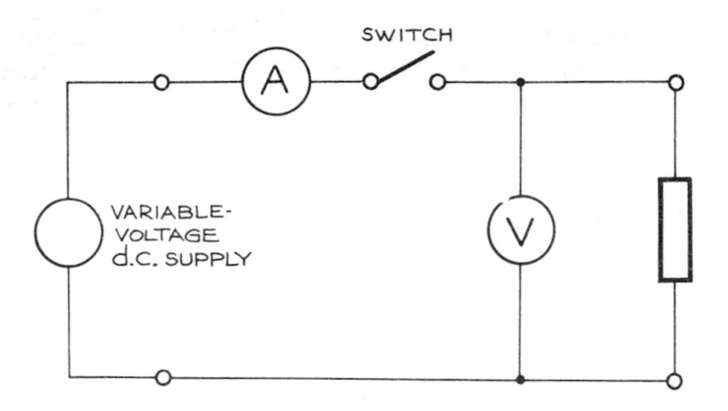

Fig. 10.12 Circuit to investigate the relationship between current and potential difference for a resistor

Notes on method

1 Connect the circuit as shown in fig. 10.12, using first the wire-wound resistor.
2 Set the variable-voltage supply to zero and close the switch.
3 Now increase the potential difference, taking regular readings of current and p.d. up to about 2 A.
4 Plot graphs of current (vertically) against potential difference horizontally.
5 Repeat using the car side-light bulb.

Results

The readings obtained are shown in Tables 10.1 and 10.2.
The graphs are shown in figs 10.13 (a) and (b).

Comments

For the wire-wound resistor, the graph of current against potential difference is a straight line with a slope of 0.2 A/V, showing a resistance of 5 Ω.

Table 10.1 Potential difference and current for a wire-wound resistor

Potential difference (volts)	Current (amperes)
1	0.22
2	0.41
4	0.83
6	1.24
8	1.61
10	2.05
12	2.42

Table 10.2 Potential difference and current for a car side-light bulb

Potential difference (volts)	Current (amperes)
0.5	0.53
1.0	0.65
2.0	0.83
3.0	0.99
4.0	1.18
5.0	1.3
9.0	1.7
12.0	2.0

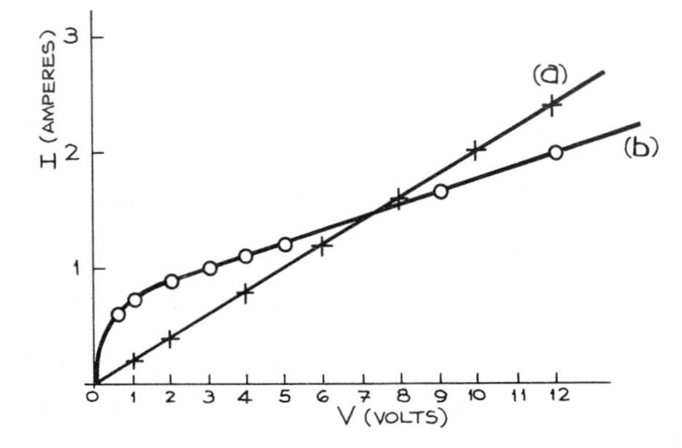

Fig. 10.13 Graphs of current against potential difference for (a) a wire-wound resistor, (b) a car side-light bulb

For the car side-light bulb the graph is non-linear due to the bulb element heating up, which causes its resistance to decrease with increased current. This effect is not evident in the case of the wire-wound resistor, since it has a power rating which ensures that negligible change in resistance occurs with temperature (see example 2 in section 11.2).

10.12 Series circuits

Consider the circuit of fig. 10.14, in which a cell is connected across three resistors which are connected one after the other. The resistors are said to be connected 'in series'. In this case a separate potential difference may be measured across each resistor as shown in fig. 10.14. The total potential difference V across the three resistors is given by adding the individual values, thus

$$V = V_1 + V_2 + V_3$$

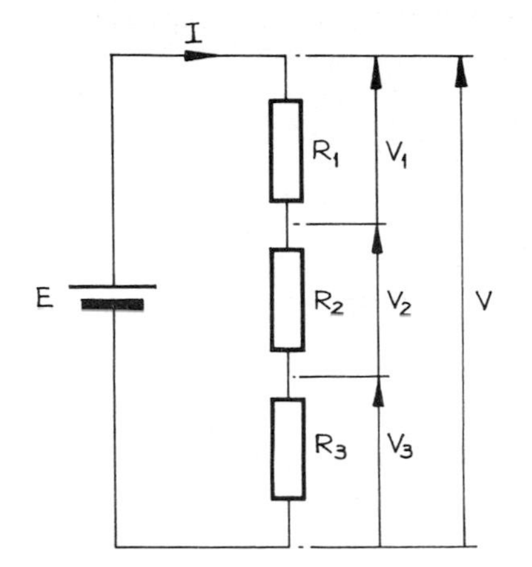

Fig. 10.14 A series-connected circuit

Since the total p.d. (V) is equal to the source e.m.f. (E) then

$$E = V = V_1 + V_2 + V_3$$

i.e. in a series circuit the sum of the potential differences is equal to the total applied e.m.f.

Notice that the same current I flows through each element of the circuit. The current flowing out of the cell is therefore the same as the current flowing in R_1, R_2, and R_3 – i.e. in a series circuit the current is the same in all parts of the circuit.

Example In the circuit of fig. 10.15, a 12 V battery is connected across the resistors R_1 and R_2. If 8 V appears across R_1, calculate (a) the potential difference that appears across R_2, (b) the current in R_2 if the resistance of R_2 is 100 Ω.

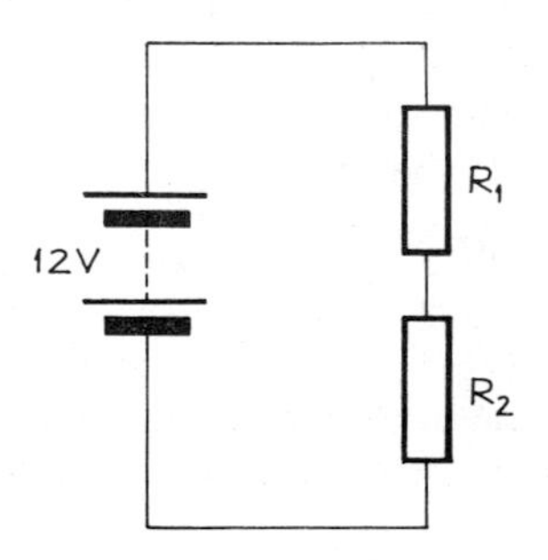

Fig. 10.15

a) p.d. across R_2 = 12 V − 8 V

= 4 V

i.e. the p.d. across R_2 is 4 V.

b) $I = \dfrac{V}{R}$

where V = 4 V and R = 100 Ω

$\therefore \quad I = \dfrac{4\text{V}}{100\,\Omega}$

= 0.04 A or 40 mA

i.e. the current is 40 mA.

10.13 Resistors in series

When resistors are connected in series as shown in fig. 10.16, then we can prove that the resultant resistance is equal to the sum of the separate values – i.e. the total resistance R_T is given by

$$R_T = R_1 + R_2 + R_3$$

The proof of this equation is as follows.

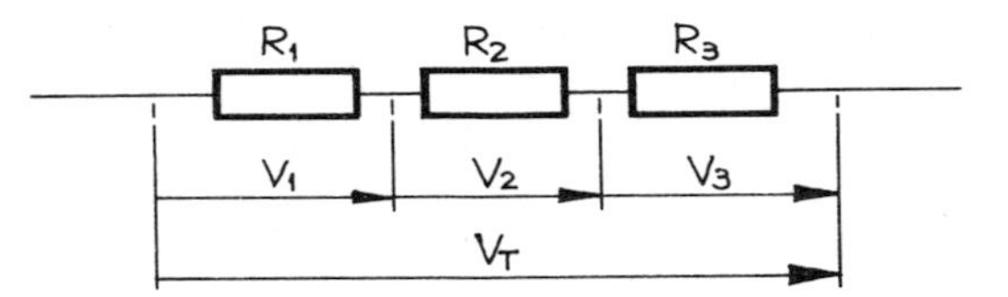

Fig. 10.16 Resistors connected in series

Referring to fig. 10.16, we have that the p.d. across R_1 is V_1, that across R_2 is V_2, and that across R_3 is V_3. Also, the total p.d. is V_T.

Since there is only one path for the current to flow, then the same current I flows in each resistor and therefore

$$V_1 = IR_1 \qquad V_2 = IR_2 \qquad V_3 = IR_3$$

and also $V_T = IR_T$

Now in a series circuit the sum of the p.d.'s is equal to the total applied p.d.

i.e. $V_T = V_1 + V_2 + V_3$

$\therefore \quad IR_T = IR_1 + IR_2 + IR_3$

Dividing throughout by I gives

$$R_T = R_1 + R_2 + R_3$$

If we have a circuit such as that shown in fig. 10.16 where the values of p.d. and resistance are known and we wish to find the current, we may apply the rule for resistors in series and say

$$I = \frac{V_T}{R_T} = \frac{V_T}{R_1 + R_2 + R_3}$$

Example Three lamps, each of resistance 500 Ω, are connected in series across a 240 V supply as shown in fig. 10.17. Calculate (a) the current in each lamp, (b) the p.d. across each lamp.

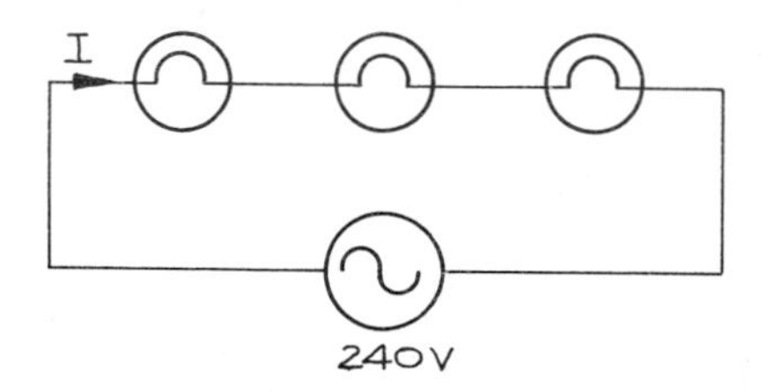

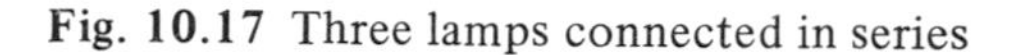
Fig. 10.17 Three lamps connected in series

a) The total resistance is given by

$$R_T = 3R$$

where $R = 500\,\Omega$

$$\therefore \quad R_T = 3 \times 500\,\Omega = 1500\,\Omega$$

$$\therefore \quad I = \frac{V}{R_T}$$

$$= \frac{240\,\text{V}}{1500\,\Omega} = 0.16\,\text{A}$$

i.e. a current of 0.16 A flows in each lamp.

b) The p.d. across each lamp is given by

$$V = IR$$

where $I = 0.16\,\text{A}$ and $R = 500\,\Omega$

$$\therefore \quad V = 0.16\,\text{A} \times 500\,\Omega = 80\,\text{V}$$

i.e. the p.d. across each lamp is 80 V. (Since the three lamps each have the same resistance, we could also have arrived at this solution by noting that one third of the total voltage exists across each lamp, i.e. 240 V/3 = 80 V.)

10.14 Parallel circuits

Figure 10.18 shows a junction in a circuit where several connections have been made. This is referred to as a 'node'.

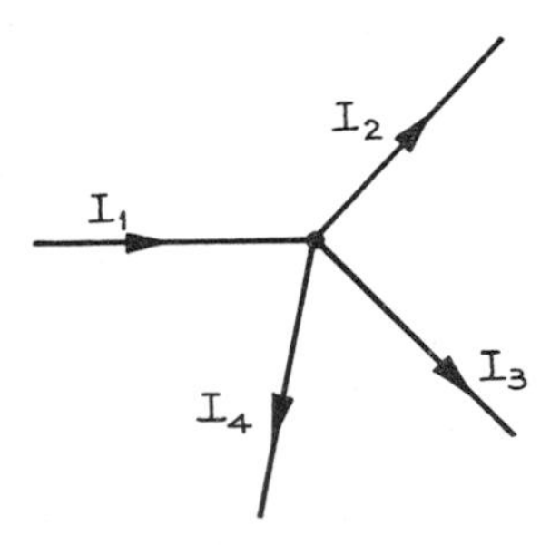

Fig. 10.18 A junction in a circuit

In the case shown in fig. 10.18, current I_1 flows into the node and currents I_2, I_3, and I_4 flow out of the node. The relationship between these currents is that

$$I_1 = I_2 + I_3 + I_4$$

i.e. at a junction in an electric circuit, the sum of the currents flowing towards the junction is equal to the sum of the currents flowing away from the junction.

Consider the circuit of fig. 10.19 in which three resistors are connected side by side with each other. They are said to be connected 'in parallel'. A current I_T flows out of the cell, but at the first junction the current splits into three so that a separate current flows in each resistor.

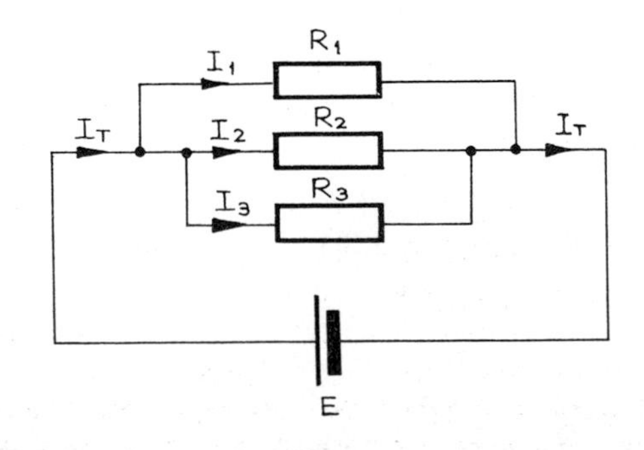

Fig. 10.19 Resistors connected in parallel

The total current I_T is given by

$$I_T = I_1 + I_2 + I_3$$

Notice that at the second junction the three currents recombine to reform the resultant current I_T.

The parallel electric circuit may be compared to a fluid system which has pipes branching off from a main pipe and then joining together again. The fluid flow would separate and then recombine just like current in the parallel electric circuit.

Example Three lamps, each of resistance 500 Ω, are connected in parallel across a 240V supply as shown in fig. 10.20. Calculate (a) the current in each lamp, (b) the total current.

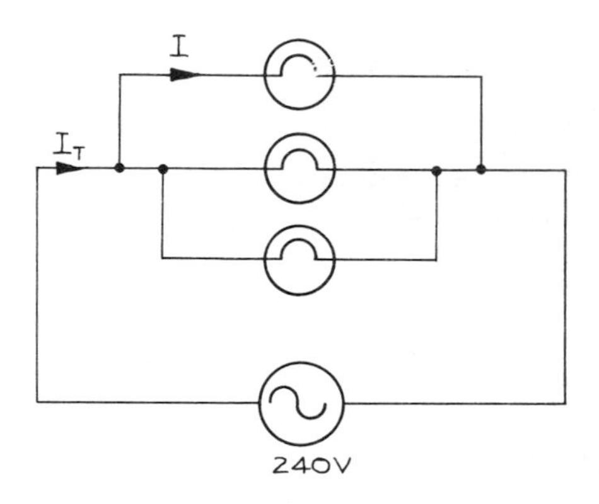

Fig. 10.20 Three lamps connected in parallel

a) The current in each lamp is given by

$$I = \frac{V}{R}$$

where $V = 240\text{V}$ and $R = 500\Omega$

$$\therefore \quad I = \frac{240\text{V}}{500\Omega} = 0.48\text{A}$$

i.e. a current of 0.48 A flows in each lamp.

b) The total current is given by

$$I_T = 3I$$

$$= 3 \times 0.48\text{A} = 1.44\text{A}$$

i.e. the total current is 1.44 A.

10.15 Resistors in parallel

When resistors are connected in parallel as shown in fig. 10.19, their total resistance R_T is given by

$$\frac{1}{R_T} = \frac{1}{R_1} + \frac{1}{R_2} + \frac{1}{R_3}$$

The proof of this equation is as follows.

Referring to fig. 10.19, the current in R_1 is I_1, that in R_2 is I_2, and that in R_3 is I_3. Also, the total current is I_T.

Since the resistors are connected in parallel, then the potential difference V across each resistor is the same; therefore

$$I_1 = \frac{V}{R_1} \qquad I_2 = \frac{V}{R_2} \qquad I_3 = \frac{V}{R_3}$$

and also $I_T = \dfrac{V}{R_T}$

Now the current flowing towards a junction is equal to the current flowing away from the junction,

i.e. $I_T = I_1 + I_2 + I_3$

$$\therefore \quad \frac{V}{R_T} = \frac{V}{R_1} + \frac{V}{R_2} + \frac{V}{R_3}$$

Dividing throughout by V gives

$$\frac{1}{R_T} = \frac{1}{R_1} + \frac{1}{R_2} + \frac{1}{R_3}$$

It is worth noting that the resultant resistance R_T is always less than the value of the smallest resistor in the combination.

Example 1 Three lamps of resistances 400 Ω, 800 Ω, and 1200 Ω are connected in parallel across a 240 V supply. Calculate (a) the resistance of the parallel combination, (b) the total current flow.

a) $$\frac{1}{R_T} = \frac{1}{R_1} + \frac{1}{R_2} + \frac{1}{R_3}$$

where $R_1 = 400\,\Omega \qquad R_2 = 800\,\Omega$ and $R_3 = 1200\,\Omega$

$$\therefore \quad \frac{1}{R_T} = \frac{1}{400\,\Omega} + \frac{1}{800\,\Omega} + \frac{1}{1200\,\Omega}$$

$$= \frac{6+3+2}{2400\,\Omega} = \frac{11}{2400\,\Omega}$$

$$\therefore \quad R_T = \frac{2400\,\Omega}{11} \approx 218\,\Omega$$

i.e. the total resistance is 218 Ω.

b) The total current is given by

$$I = \frac{V}{R_T}$$

$$= \frac{240\,\text{V}}{218\,\Omega} = 1.1\,\text{A}$$

i.e. the total current is 1.1 A.

Example 2 Three identical 240 V bulbs are to be connected in a circuit across a 240 V supply. They are tried first as a series circuit and then as a parallel circuit. State (a) in which case the bulbs would be brightest, (b) in which case failure of one of the bulbs due to an open circuit would cause all of the bulbs to go out.

a) Parallel (see the examples in sections 10.13 and 10.14).
b) Series.

The case of two resistors in parallel is worth considering in more detail.

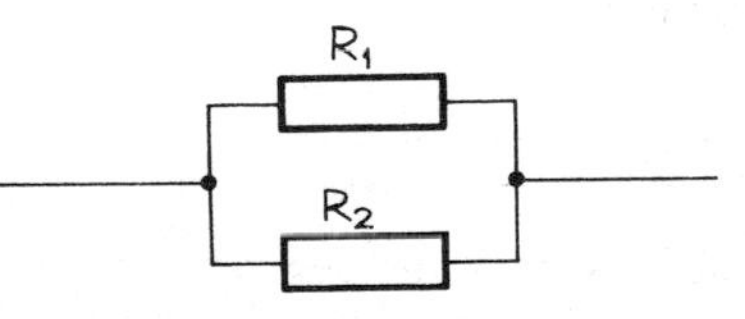

Fig. 10.21 Two resistors in parallel

For the arrangement shown in fig. 10.21,

$$\frac{1}{R_T} = \frac{1}{R_1} + \frac{1}{R_2}$$

$$= \frac{R_1 + R_2}{R_1\, R_2}$$

$$\therefore \quad R_T = \frac{R_1\, R_2}{R_1 + R_2} = \frac{\text{product}}{\text{sum}}$$

This equation is often used and is therefore worth remembering.

Example 3 Calculate the total resistance of two 470 Ω resistors connected in parallel.

$$R = \frac{\text{product}}{\text{sum}}$$

$$= \frac{470\,\Omega \times 470\,\Omega}{470\,\Omega + 470\,\Omega}$$

$$= 235\,\Omega$$

i.e. the resultant of two 470 Ω resistors connected in parallel is 235 Ω.

10.16 Resistivity

Samples of different materials having the same dimensions possess different resistances to the flow of electric current.

We have discussed previously that some materials have a very low resistance and are referred to as conductors - examples are copper and aluminium. Other materials have a very high resistance and are referred to as insulators - examples are mica, paper, and p.v.c.

To make a comparison of materials as to their resistance, it is of course necessary to compare samples with the same dimensions. For example, an iron wire of given length and diameter has about 7.5 times the resistance of a similar copper wire, while a Eureka wire of the same dimensions has 30 times the resistance of the copper wire. (Eureka is an alloy (a mixture) of copper and nickel.)

This property of materials which gives them different resistances for the same dimensions is referred to as *resistivity*.

Resistivity is a constant for a given material and is given the symbol ρ (the Greek letter *rho*).

The resistance of a conductor also depends on the dimensions of that conductor. The resistance is proportional to the length of the conductor; so that the longer the wire, the larger is its resistance.

Also, resistance decreases with increased cross-sectional area, since it is easier for an electric current to flow in a thick wire than in a thin one; i.e. the resistance is inversely proportional to the cross-sectional area.

If we have a wire of length l and cross-sectional area a, then the resistance of the wire is given by

$$R = \text{constant} \times \frac{\text{length}}{\text{cross-sectional area}}$$

$$= \text{constant} \times \frac{l}{a}$$

and this constant is the resistivity ρ.

$$\therefore \quad R = \rho \frac{l}{a}$$

We may rearrange the equation above to make ρ the subject of the equation:

$$\rho = \frac{Ra}{l}$$

where R is in ohms (Ω), a is in metres squared (m^2), and l is in metres (m). Thus the units of resistivity are ohm metres squared per metre, or more simply ohm metres (Ωm). The unit $\mu\Omega$m is also used.

The resistivity of a material may thus be formally defined as the resistance of a specimen of the material one metre long and with a cross-sectional area of one metre squared.

This is of course an extremely large block of material, which it would not be practical to use in order to measure resistivity. The large cross-sectional area used in the definition also means that the resistance of this unit cube is very small. For example, the resistivity of annealed copper at 20°C is 0.00000001725 Ωm, which means that a 1 metre cube would have a resistance between two faces of $0.01725 \times 10^{-6}\ \Omega$ or $0.01725\ \mu\Omega$.

In practice it is more sensible to take a measured length of wire of uniform cross-section, measure its resistance, and obtain the resistivity from the equation

$$\rho = \frac{Ra}{l}$$

Example 1 The resistivity of copper is $0.017\ \mu\Omega$ m. Calculate the resistance of 30 m of copper wire of cross-sectional area 1.5 mm.

$$R = \rho\frac{l}{a}$$

where $\rho = 0.017 \times 10^{-6}\ \Omega\text{m} \qquad l = 30\text{m}$

and $a = 1.5\text{mm}^2 = 1.5 \times 10^{-6}\ \text{m}^2$

$$\therefore \quad R = 0.017 \times 10^{-6}\ \Omega\text{m} \times \frac{30\text{m}}{1.5 \times 10^{-6}\ \text{m}^2}$$

$$= 0.34\Omega$$

i.e. the resistance of a 30 m length of copper wire of cross-sectional area 1.5 mm is 0.34Ω.

Example 2 A piece of aluminium wire has a uniform cross-sectional area of 4mm^2. If a 100 m length has a resistance of $0.675\ \Omega$, calculate the resistivity of aluminium.

$$\rho = \frac{Ra}{l}$$

where $R = 0.675\ \Omega \qquad a = 4\ \text{mm}^2 = 4 \times 10^{-6}\ \text{m}^2$ and $l = 100\ \text{m}$

$$\therefore \quad \rho = \frac{0.675\,\Omega \times 4 \times 10^{-6}\,\text{m}^2}{100\,\text{m}}$$

$$= 0.027 \times 10^{-6}\,\Omega\text{m} \quad \text{or} \quad 0.027\,\mu\Omega\text{m}$$

i.e. the resistivity of aluminium is $0.027\,\mu\Omega$m.

10.17 Effect of temperature on resistance

When a conductor is heated, its resistance increases. Similarly, cooling a conductor lowers its resistance. The reason is as follows.

All materials are made up of atoms which vibrate. In solid materials, the atoms vibrate about their resident position. As the solid is heated, the vibration of the atoms increases. In a conductor which is carrying an electric current, the flow of the electric current (i.e. the electron flow) is affected by the vibrating atoms just as if the electrons were people pushing their way through a crowd. When the conductor is heated, the increased atomic vibration makes it more difficult for the electrons to pass without banging into more vibrating atoms, thus the resistance to flow of electric current is increased.

The resistance of all pure metals such as copper, tungsten, iron, etc. increases with increase in temperature for the reason stated above.

However, be wary - in the case of insulators such as paper and plastic, or of elements such as carbon and silicon, the resistance decreases with increase in temperature. This is not to say that in these materials the atoms do not vibrate more with increasing temperature - indeed they do, and the same effect occurs as with pure metals; but there is also another effect taking place.

The other effect is that heating the material gives more electrons sufficient energy to break free from the parent atoms, and there is thus an increase in the number of charge-carrying free electrons. In insulators, this effect predominates and causes the resistance to fall.

The variation of resistance with temperature is used extensively as a means of measurement of temperature.

The curves of fig. 10.22 show the variation of resistance with temperature for platinum and for nickel. The vertical axis shows the resistance at any temperature as a proportion of the resistance at 0°C.

Nickel has the greatest variation in resistance with temperature, but the graph is non-linear and therefore nickel is not a suitable material to use as a temperature-sensor.

Platinum has a reasonably high and fairly linear resistance change with temperature and is thus the metal most commonly used in temperature-sensors (despite its high cost).

A device whose resistance varies quite markedly with temperature is the thermistor. The most common types have a very non-linear negative temperature coefficient (i.e. their resistance *decreases* with increasing temperature), as shown in fig. 10.23. They are made from oxides of iron, nickel, and cobalt with small amounts of other substances.

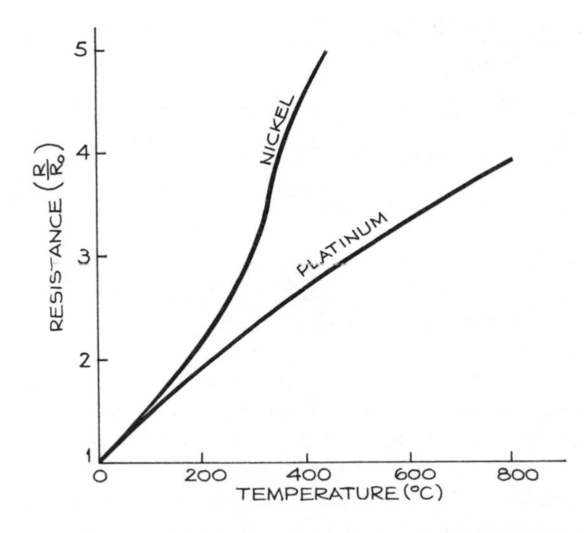

Fig. 10.22 Variation of resistance with temperature for platinum and nickel

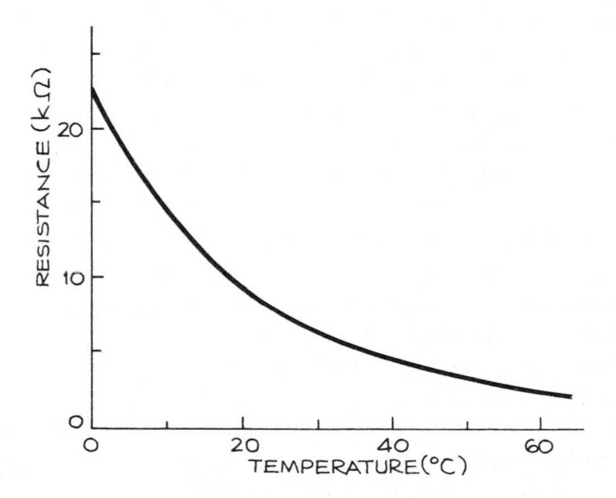

Fig. 10.23 Variation of resistance with temperature for a thermistor with a negative temperature coefficient

Some types of thermistor are encapsulated in a vacuum inside glass beads and are used in temperature-compensation in electronic circuits such as audio amplifiers, to compensate for variation of component parameters with temperature. Others are encapsulated in stainless-steel probe assemblies and are designed primarily for temperature measurement and control, flow measurement, and liquid-level detection.

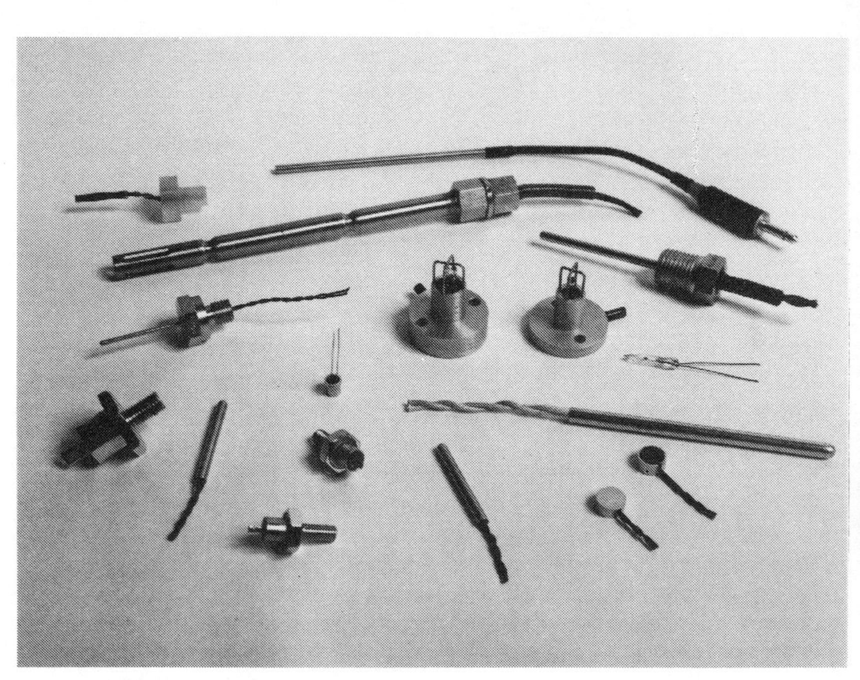

A range of industrial thermistors housed in a variety of encapsulations

Thermistors have the advantages of being very small (and therefore having a low heat capacity) and of having a large temperature coefficient.

Exercises on chapter 10

1 The current/p.d. relationship for two different resistors is as shown in fig. 10.24. Determine the value of each resistor.

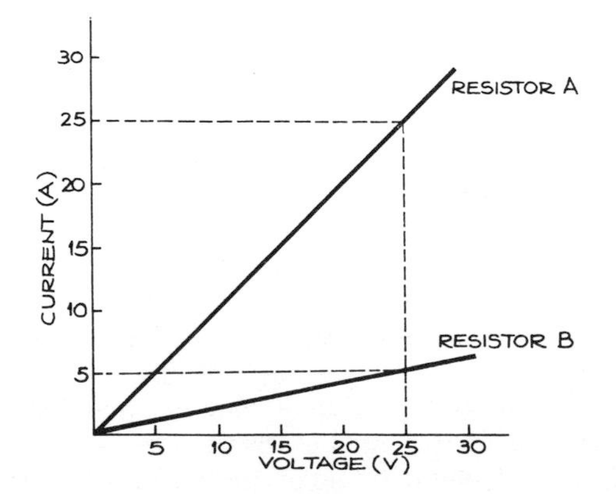

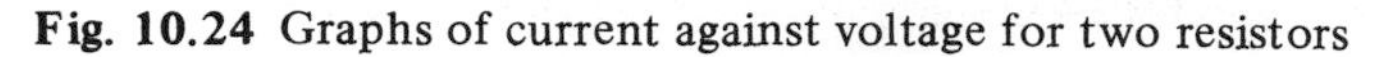

Fig. 10.24 Graphs of current against voltage for two resistors

2 A battery of negligible internal resistance is connected to a resistance of 30 Ω and a current of 1.5 A flows. (a) Determine the battery e.m.f. (b) If the original battery is now replaced by one having twice the e.m.f. and with negligible internal resistance, determine the current flow.
3 If three identical filament lamps are connected in parallel and the combined resistance is 120 Ω, calculate the resistance of one lamp.
4 A piece of copper wire has a resistance of 5 Ω. It is stretched so that its length is doubled, and its cross-sectional area decreases correspondingly. Calculate the new resistance of the copper wire. (Hint: calculate the ratio of the cross-sectional areas.)
5 (a) State Ohm's law. (b) A 12 V battery is connected across a load having a resistance of 30 Ω. Determine the current flowing in the load.
6 Two resistors are connected in series across a 20 V supply and a current of 5 A flows in the circuit. If one of the resistors has a value of 1 Ω, determine (a) the value of the other resistor, (b) the potential difference across the 1 Ω resistor.
7 For the circuit shown in fig. 10.25, determine (a) the value of the current I, (b) the value of the resistor R.

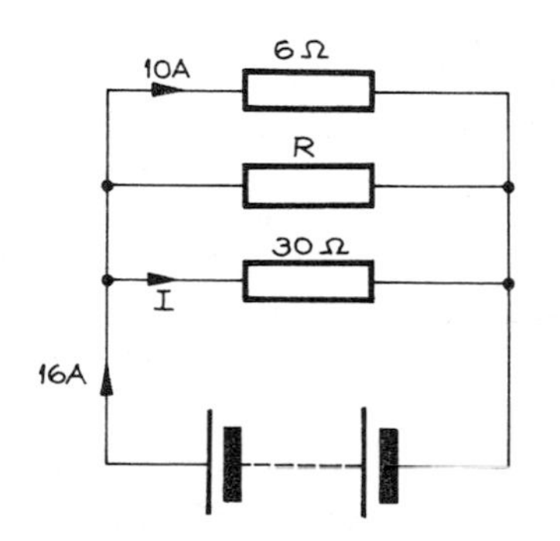

Fig. 10.25 A parallel electric circuit

8 The resistance of a 1 km length of wire of cross-sectional area 0.17 mm is 100 Ω. Determine the resistivity of the wire in μΩ m units.
9 Explain why platinum is used as the temperature-sensitive element in a device for measuring temperature. State one advantage and one disadvantage of using a thermistor.
10 State what are meant by the terms (a) a good insulator, (b) a good conductor. Give two examples of each.
11 State three effects of an electric current and give two examples of the application of each.
12 Obtain a circuit diagram for a piece of electronic equipment (say a radio or a minicomputer) and identify each component.
13 State the units in which the following quantities are measured: (a) resistance, (b) potential difference, (c) current, (d) e.m.f.

14 Calculate the current which flows when 240 V is connected across a 1.5 kΩ resistor.
15 A resistor has 15 V connected across it and a current of 2.2 mA flows. Calculate the resistance.
16 An electric-fire element has a resistance of 65 Ω. Calculate the current flow when 240 V is connected across it.
17 An electric light-bulb takes a current of 0.25 A when connected across a 240 V supply. Calculate the resistance of the bulb.
18 A cable of resistance 0.1 Ω carries a current of 20 A. Calculate the potential difference between the ends of the cable.
19 Two 570 Ω resistors placed in series are connected across a 440 V supply. Calculate the current.
20 Two 20 Ω resistors are connected in parallel and this group is then connected in series with a 4 Ω resistor. Calculate the total effective resistance of the circuit.
21 An experiment to find the resistance of a coil gave the following readings on an ammeter and voltmeter:

Potential difference (V)	0	1.1	1.95	3.15	3.9
Current (A)	0	0.5	1.0	1.5	2.0

Plot a graph of p.d. against current and hence find the resistance of the coil.
22 When the four identical hotplates on a cooker are all in use, the current taken from a 240 V supply is 33.3 A. Calculate (a), the resistance of each hotplate, (b) the current taken when only three hotplates are switched on. The hotplates are connected in parallel.
23 Calculate the total current when six 120 Ω torch bulbs are connected in parallel across a 9 V supply of negligible internal resistance.
24 When two identical fans are connected in series across a 240 V supply, the total current is 0.52 A. Calculate (a) the potential difference across each fan, (b) the resistance of each fan, (c) the current taken if the two fans are connected in parallel across the supply.
25 An electric kettle takes 12.5 A from a 240 V supply. Calculate the current that would flow if the kettle were connected across a 110 V supply.
26 An electric shaver takes 0.5 A from a 110 V supply. Calculate the resistance to be connected in series so that the combination takes the same current from a 240 V supply.
27 Two lathes connected in parallel take 12 A from a 240 V supply. If a milling machine with an electrical resistance of 13.3 Ω is also connected in parallel, calculate (a) the resistance of the total parallel combination, (b) the total current taken from the supply.

11 Power in electric circuits

11.1 Energy in electric circuits

When an electric current flows through a conductor, heat is generated and electrical energy is thus converted into heat energy.

The laws relating to the heating effect of an electric current were discovered by James P. Joule (1818–1889) whose name is commemorated in the unit of energy, the joule (symbol J).

He found that the heat generated in a conductor is proportional to

a) the square of the current – i.e. if the current is doubled, the heat generated increases fourfold;
b) the resistance of the conductor;
c) the time for which the current flows.

Thus the heat energy W generated by a current I flowing in a resistance R for a time t is given by

$$W = I^2Rt$$

and when $I = 1$ ampere, $R = 1$ ohm, and $t = 1$ second then $W = 1$ joule (1J).

Example A cooker element has a resistance of 30 Ω and is fed from a 240 V mains supply. Calculate (a) the current, (b) the energy consumed in one hour.

a) $I = \dfrac{V}{R}$

where $V = 240\text{V}$ and $R = 30\Omega$

$\therefore \quad I = \dfrac{240\text{V}}{30\Omega} = 8\,\text{A}$

i.e. the current is 8 A.

b) $W = I^2Rt$

where $I = 8\,\text{A}$ $\quad R = 30\Omega$ and $t = 3600\text{s}$

$$\therefore \quad W = (8\,\text{A})^2 \times 30\Omega \times 3600\,\text{s}$$

$$= 6912000\text{J}$$

$$= 6.912 \times 10^6\,\text{J} \quad \text{or} \quad 6.912\,\text{MJ}$$

i.e. the energy consumed in one hour is 6.912 MJ.

11.2 Power in electric circuits

Power is the rate of energy transfer or conversion. In electric circuits, power is the rate at which electrical energy is converted into other forms of energy, such as heat.

The power is thus the total energy conversion divided by the time taken for the conversion.

i.e. $$\text{power} = \frac{\text{energy converted in joules}}{\text{time in seconds}}$$

or $$P = \frac{W}{t} = \frac{I^2Rt}{t} = I^2R$$

The unit of power is the watt (symbol W), named after James Watt (1736–1819), and 1 watt is 1 joule per second.

Two alternative expressions for power may be derived. Since $R = V/I$, we may write

$$P = I^2R = I^2 \times \frac{V}{I}$$

$$= VI$$

Also, since $I = V/R$, we may write

$$P = (I)^2R = \left(\frac{V}{R}\right)^2 R$$

$$= \frac{V^2}{R}$$

Hence we have

$$P = I^2R = IV = \frac{V^2}{R}$$

These three equations may be used to calculate the power in d.c. electric circuits. (In alternating-current circuits a set of modified equations is required.)

Electrical engineers tend to memorise the expression for power $P = VI$ as

watts = volts × amperes

Example 1 A 40 W bulb is connected to a 240 V mains supply. Calculate (a) the current taken, (b) the resistance of the bulb filament.

a) $P = IV$

$\therefore \quad I = \frac{P}{V}$

where $P = 40\text{W}$ and $V = 240\text{V}$

$$\therefore \quad I = \frac{40\,\text{W}}{240\,\text{V}}$$

$$= 0.17\,\text{A}$$

i.e. the current taken is 0.17 A.

b) $R = \dfrac{V}{I}$

where $V = 240\,\text{V}$ and $I = 0.17\,\text{A}$

$$\therefore \quad R = \frac{240\,\text{V}}{0.17\,\text{A}}$$

$$= 1412\,\Omega \quad \text{or} \quad 1.4\,\text{k}\Omega$$

i.e. the resistance is 1.4 kΩ.

Example 2 A 680 Ω carbon resistor to be used in an electronic circuit is rated at $\frac{1}{2}$ watt. Calculate the maximum d.c. potential difference that may be applied across the resistor without overheating.

$$P = \frac{V^2}{R}$$

$$\therefore \quad V = \sqrt{PR}$$

where $P = 0.5\,\text{W}$ and $R = 680\,\Omega$

$$\therefore \quad V = \sqrt{0.5\,\text{W} \times 680\,\Omega}$$

$$= \sqrt{340}\,\text{V}$$

$$= 18.4\,\text{V}$$

i.e. the maximum potential difference that may be applied across a 680 Ω $\frac{1}{2}$ watt resistor is 18.4 V.

Example 3 Calculate the power supplied by a 200 V d.c. generator when feeding a current of 12 A.

$$P = VI$$

where $V = 200\,\text{V}$ and $I = 12\,\text{A}$

$$\therefore \quad P = 200\,\text{V} \times 12\,\text{A}$$

$$= 2400\,\text{W} \quad \text{or} \quad 2.4\,\text{kW}$$

i.e. the power supplied is 2.4 kW

Example 4 A 2 kW electric heater takes a current of 10 A. Calculate its resistance.

$$P = I^2R$$

$$\therefore \quad R = \frac{P}{I^2}$$

where $P = 2\,\text{kW} = 2000\,\text{W}$ and $I = 10\,\text{A}$

$$\therefore \quad R = \frac{2000\,\text{W}}{(10\,\text{A})^2}$$

$$= 20\,\Omega$$

i.e. the resistance is $20\,\Omega$.

Example 5 Calculate the power dissipated by a $100\,\Omega$ resistor connected across an e.m.f. of 50V.

$$P = \frac{V^2}{R}$$

where $V = 50\,\text{V}$ and $R = 100\,\Omega$

$$\therefore \quad P = \frac{(50\,\text{V})^2}{100\,\Omega}$$

$$= 25\,\text{W}$$

i.e. the power dissipated is 25W.

11.3 The kilowatt hour

A joule is a fairly small quantity of energy, and the unit is not very appropriate for measuring the energy consumption of domestic electric appliances, say, since we end up with some large numbers (as in the example in section 11.1).

A more appropriate alternative unit of energy which is often used by electrical engineers is the kilowatt hour (abbreviation kWh). This is the quantity of energy converted into some other form when one kilowatt is consumed for one hour,

i.e. $1\,\text{kWh} = 1\,\text{kW} \times 1\,\text{h}$

This unit is commonly used when calculating the quantity of electrical energy supplied to a domestic or industrial consumer by the electricity board, i.e. it is the unit used in domestic and industrial tariffs.

Example 1 If a cooker element is rated at 2kW, calculate the energy consumed (in kWh) in one hour.

$$W = 2\,\text{kW} \times 1\,\text{h}$$

$$= 2\,\text{kWh}$$

i.e. the energy consumed by the element in one hour is 2kWh.

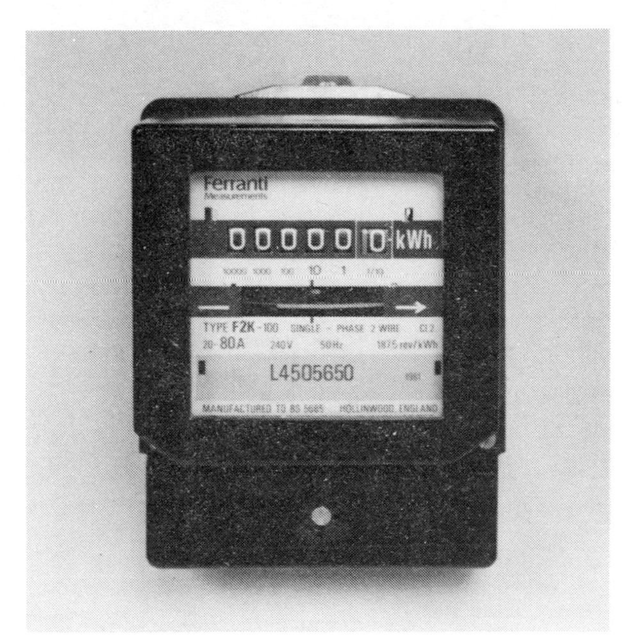

A kW h domestic tariff meter

Example 2 Three 60 W lamps and four 100 W lamps are run for 5 hours each day for 1 week (7 days). Calculate (a) the total number of kilowatt hours consumed, (b) the total cost if electrical energy costs 5p per unit.

a) Total power $= 3 \times 60\,\text{W} + 4 \times 100\,\text{W}$

$= 180\,\text{W} + 400\,\text{W}$

$= 580\,\text{W}$ or $0.58\,\text{kW}$

Total kWh consumed $= 0.58\,\text{kW} \times 5\,\text{h/day} \times 7$ days

$= 20.3\,\text{kWh}$

b) Total cost $= 20.3\,\text{kWh} \times 5\,\text{p}$

$= 101.5\,\text{p}$

i.e. the total energy consumed by the lamps in a week is 20.3 kWh at a cost of about £1.

11.4 Fuses

A fuse is included in an electric circuit as a safety precaution. It contains simply a wire which melts when it passes a current greater than that for which it is rated. Fuses are used with equipments which are connected to the mains 240 V supply, as a protection against electric shock.

The outer case of a metal appliance such as a kettle is earthed (i.e. connected to the earth pin on the plug), and the fuse is connected in series with the live wire of the supply as shown in fig. 11.1(a). If a short circuit (a low-resistance fault) occurs between the live wire and the earth, due for example to corrosion of the kettle element, then the much increased current flow will cause the fuse to 'blow' (go open circuit). This will effectively disconnect the appliance from the supply and make it impossible for the metal case to become live.

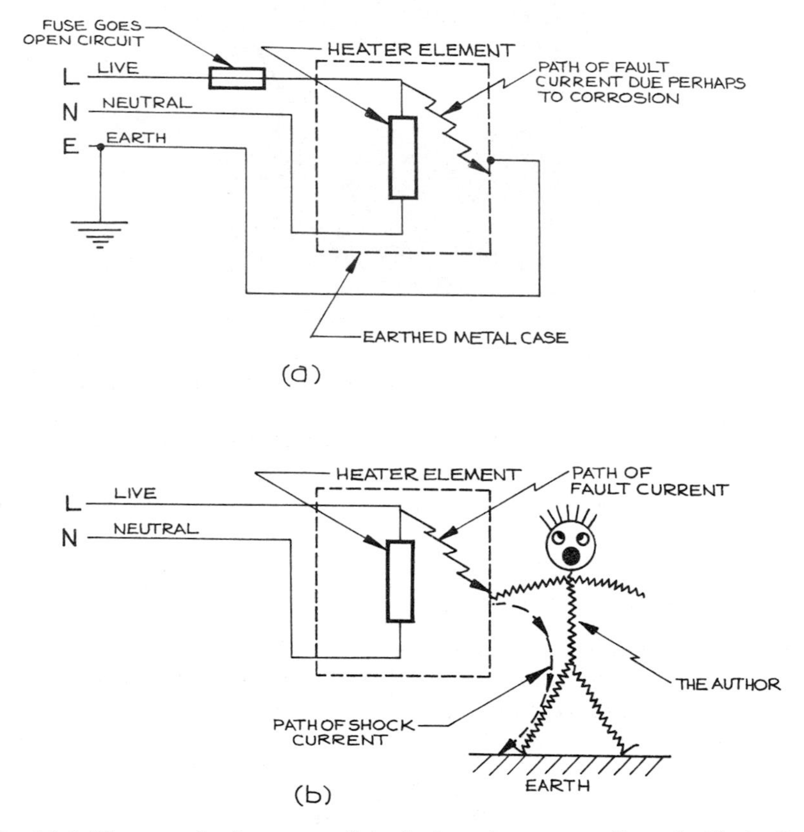

Fig. 11.1 The use of a fuse as a safety device when connecting a kettle to the mains (the fuse is included in the mains plug)

If it were not for the fuse, a fault of this kind could lead to electric shock, as shown in fig. 11.1(b).

The 'earth' in a consumer electric circuit is generally a large metal spike driven into the ground close to the building and connected to the earth wire on the three-wire supply.

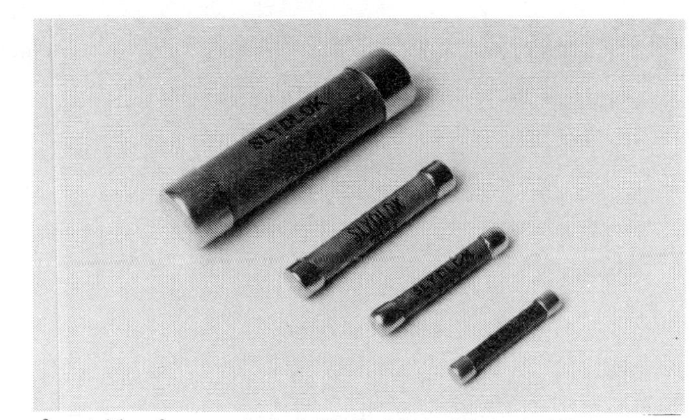

A range of cartridge fuses, rated at 60 A, 30 A, 15 A, and 5 A

11.5 Wiring a three-pin plug

Most domestic appliances are fitted with a three-pin plug referred to as a '13 amp' plug. This indicates the maximum fuse rating for which this plug is suitable - not the only fuse rating. In fact the plug should be fitted with a fuse which is rated most appropriately for the appliance being used. For example, a 100W bulb in a table lamp will take approximately 0.5 A and the plug should therefore be fitted with say a 1 A fuse.

The plug is fitted to a standard three-core flexible lead using the following convention:

live pin - brown wire

neutral pin - blue wire

earth pin - green/yellow wire

The plug should thus be wired as shown in fig. 11.2.

The cord grip is used to ensure a good mechanical support, so that strain is not put on the electrical connections.

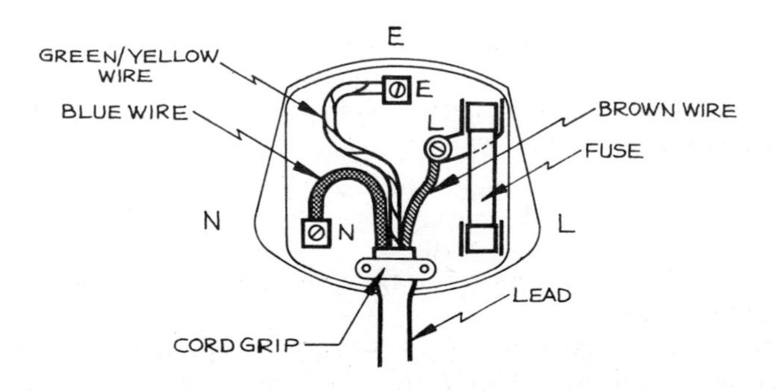

Fig. 11.2 Wiring a '13 amp' plug

Example A television set has a power rating of 200 W. Suggest a suitably rated fuse to be used in connecting the appliance to a 240 V mains supply, if the fuses available are 2 A, 3 A, 5 A, and 13 A.

$$P = VI$$

$$\therefore \quad I = \frac{P}{V}$$

where $P = 200\text{W}$ and $V = 240\text{V}$

$$\therefore \quad I = \frac{200\text{W}}{240\text{V}} = 0.83\text{A}$$

i.e. a current of 0.83 amperes will flow in the appliance. A suitable fuse would be one with a current rating a little higher than the expected current, i.e. a 2 A fuse in this case.

Exercises on chapter 11

1 An electric light-bulb has a rating of 100 W. Assuming that the only useful form of energy output is light (heat and other forms of energy being wasted), calculate the effective efficiency of the bulb if in a 20 s period it delivers 400 J of light energy.
2 Determine the power rating of the element of an electric fire of resistance 25 Ω if a current of 10 A flows through it.
3 If fuses rated at 5 A, 10 A, and 13 A are available, state which would be most appropriate for an electric fire rated at 2 kW 240 V.
4 Calculate the power dissipated when a current of 5 mA flows through a resistance of 4 kΩ.
5 A 240 V supply feeds a current of 2 A for 5 h. Calculate the energy used.
6 Calculate the supply e.m.f. if 1.5 kJ of energy is used in 30 s when a current of 5 A flows in a resistor.
7 Calculate the energy used in one hour when an 8.3 Ω resistor is supplied from 240 V.
8 A current of 5 A flows for 10 min between two points at a potential difference of 24 V. Calculate (a) the energy converted in kW h, (b) the cost of the energy at 5p per kW h.
9 A 3 kW heater is switched on for 1.5 h. Calculate the energy converted, in kW h and in joules. Hence state the relationship between kW h and joules.
10 Name the SI unit of electrical energy. The cost of operating two 3 kW electric heaters for 8 h is 120p. Calculate the cost of energy per kW h.
11 A 3 kW immersion heater is designed to operate from a 250 V supply. Calculate its resistance and the current taken by the heater when operating at its rated potential difference. If the heater is now used on a 200 V supply, what will be the power dissipated, assuming that the resistance remains the same?
12 Calculate the cost of operating a 5 kW electric motor for 12 hours if energy is charged at 5p per kW h. If the supply e.m.f. is 400 V, calculate the supply current.

12 Chemical effects of an electric current

12.1 Introduction

As we shall see in the next section, there is a large group of solid compounds which when dissolved in water separate into what are called positive and negative *ions*, the resulting solution being referred to as an *electrolyte*.

The study of the various effects and uses of electrolytes is referred to as *electrochemistry* and is important both in the electroplating industry and to electric-cell manufacturers.

12.2 Ions and electrolytes

Atoms and molecules are electrically neutral, but atoms can lose or gain electrons and thus become charged particles known as ions.

Certain molecules can split up into two or more charged units, but always such that the total number of positive charges is equal to the total number of negative charges. Such charged units are also referred to as *ions*.

There is a large group of solid compounds which when dissolved in water separate into positive and negative ions, the solution being referred to as an *electrolyte*. A typical example is sodium chloride (common salt). When sodium chloride is dissolved in water, the sodium atom in each molecule of sodium chloride leaves its one outer electron with the chlorine atom. The positive charge on the sodium nucleus is now greater than the total negative charge on the surrounding electrons, so a positively charged sodium ion is formed. Similarly, the total negative charge on the electrons surrounding the chlorine nucleus is now greater than the positive charge on the nucleus, so a negative ion is formed which is referred to as a chloride ion.

As each molecule forms one ion with a single positive charge and one ion with a single negative charge, the total positive and negative charges in the solution are balanced and the solution is electrically neutral.

The reaction is shown by the equation

$$NaCl \rightleftharpoons Na^+ + Cl^-$$

This represents an equilibrium situation in which sodium chloride is continually breaking up into its ions, while other sodium and chloride ions are continually recombining to form sodium chloride.

The electric charge on an ion depends on the number of electrons in the outer shell of the original atom. For example, the iron atom has two outer-shell electrons, and one iron atom combines with two chlorine atoms to form one molecule of ferrous chloride ($FeCl_2$).

When ferrous chloride is dissolved in water, each molecule forms one ferrous ion with two positive charges and two chloride ions each with a single negative charge:

$$FeCl_2 \rightleftharpoons Fe^{++} + 2Cl^-$$

When copper sulphate ($CuSO_4$) dissolves in water, each molecule separates into one copper ion with two positive charges (since the copper atom has two outer-shell electrons) and one sulphate ion with a double negative charge:

$$CuSO_4 \rightleftharpoons Cu^{++} + SO_4{}^{--}$$

All such positive and negative ions in solution tend to become attached to water molecules, and in this condition they can move independently and freely within the solution. This ionic mobility means that such solutions can conduct electricity - the electric current in this case being a drift of positive charge carriers (ions) in one direction as well as a drift of negative charge carriers (ions) in the opposite direction.

12.3 Electroplating

Figure 12.1 shows a solution of copper sulphate ($CuSO_4$) dissolved in water in a container. We have seen that charged copper and sulphate ions exist in the solution.

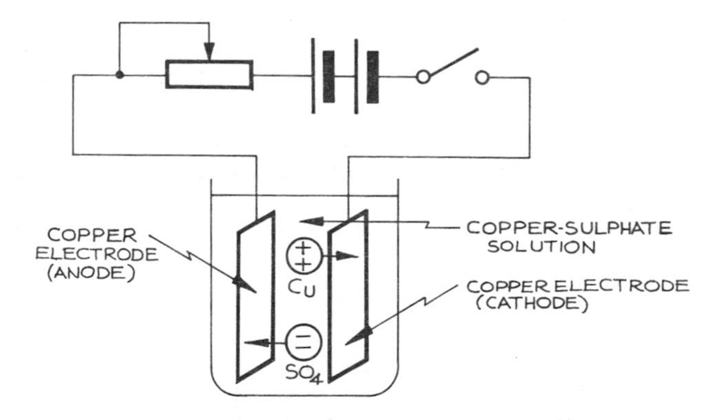

Fig. 12.1 Electroplating using copper-sulphate solution

Two copper electrodes are inserted into the solution, and a cell is connected across them in series with a variable resistor and a switch, as shown in fig. 12.1.

When the switch is closed, the positively charged copper ions move towards the negative electrode (the cathode). At the cathode, each copper ion absorbs two electrons to form a neutral atom of copper which is deposited on to the surface of the copper plate.

The negatively charged sulphate ions move towards the positive electrode (the anode) where they each give up two electrons to form neutral SO_4. This cannot exist unchanged and combines with atoms of the copper electrodes to form copper sulphate which passes back into the solution. The density of the electrolyte thus remains the same.

The net effect is to transfer copper from the anode to the cathode. This is the basis of the electroplating industry.

It is not necessary for the two electrodes to be made of the same element, so this phenomenon can be used to electroplate metals such as steel with a more corrosion-resistance metal such as chromium. In this case the anode is chromium, which is transferred to the steel cathode to form a plating which both enhances the appearance and protects the base metal from corrosion.

12.4 Electric cells

Electric cells are sources of electrical energy. When a conductor is connected across the two terminals of a cell, a current flows which is produced by electrochemical action.

The cell consists of an *electrolyte* and two *electrodes* which must be of different materials. Almost any two dissimilar metals can be used as electrodes, but only certain combinations of electrolyte and electrodes will give a useful performance in terms of current output.

The simple cell

Let us consider the action of a simple cell such as that shown in fig. 12.2.

There are two electrodes - one of copper and one of zinc. The electrolyte is a dilute solution of sulphuric acid which contains positively charged hydrogen ions and negatively charged sulphate ions.

The zinc plate has a high tendency to go into solution in the acid, forming positively charged zinc ions in the solution and leaving electrons on the plate, which therefore becomes negatively charged.

The copper plate accepts positive ions from the solution and thus accumulates a positive charge.

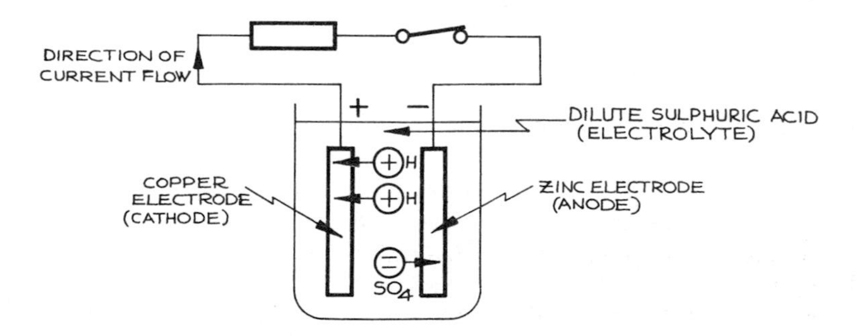

Fig. 12.2 A simple electric cell

The zinc plate is thus referred to as the *anode* and the copper plate as the *cathode* (see section 2.7).

When the switch is closed, a complete circuit is formed via the conductor and, as one plate is positive and the other negatively charged, the potential difference between the plates causes an electric current to flow.

The positively charged hydrogen ions in the electrolyte are attracted towards the copper electrode (the cathode), where they each gain an electron. Hydrogen atoms are thus formed, which bond together to form hydrogen molecules (H_2), and hydrogen bubbles are produced at the cathode.

The negatively charged sulphate ions move towards the zinc electrode (the anode) and combine with zinc atoms to form zinc sulphate which goes back into solution. This has the effect of gradually corroding the zinc electrode.

The cell will deliver current until either the zinc plate has completely dissolved or there are no more hydrogen ions left in the electrolyte.

One disadvantage of the simple cell is that the generation of gas at the cathode results in that electrode becoming completely covered with hydrogen bubbles, which makes it ineffective. This is due to two effects - firstly the bubbles act as a shield, thus increasing the resistance of the cell; secondly, the gas itself acts as an electrode, setting up a potential which is opposed to that of the metal. The overall effect is referred to as *polarisation*.

To overcome this difficulty, several types of cell contain a depolarising substance which combines with the hydrogen as it is formed.

Another disadvantage of this cell is that, due to slight impurities in the zinc, *local electrochemical action* takes place at the anode - i.e. the zinc and the impurities act as the electrodes for small cells on the anode of the main cell itself, and these small corrosion cells considerably speed up the erosion of the zinc.

Local electrochemical action can be overcome by first plating the zinc with mercury, which dissolves pure zinc out of the plate and forms a bright amalgam over the surface. Local action cannot now occur, since the amalgam prevents the impurities from coming into contact with the acid.

12.5 Primary and secondary cells

Electric cells may be classified as either primary cells or secondary cells.

Primary cells

Primary cells may not be recharged, since the conversion of chemical energy into electrical energy is not reversible with this type of cell construction.

Types of primary cell are the simple cell described in section 12.4 and the dry cell used in torches and portable radios.

The dry cell produces an e.m.f. of about 1.5 V. If a dry cell as big as a bus were constructed, it would still only yield 1.5 V but it would last a very long time.

The two major dry-cell types involved in the consumer market are the traditional zinc-carbon cell and the newer, more durable, and longer lasting alkaline-manganese cell.

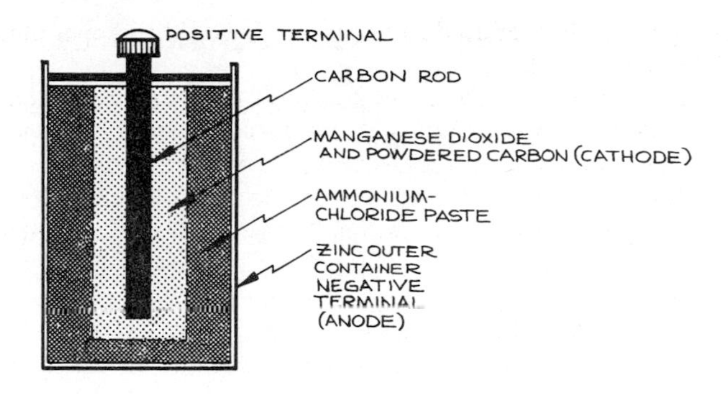

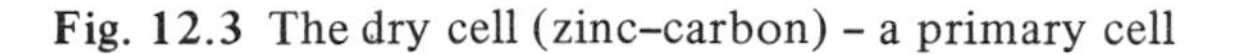

Fig. 12.3 The dry cell (zinc–carbon) – a primary cell

The zinc-carbon cell is shown in fig. 12.3. It has a zinc anode, a manganese-dioxide cathode surrounding a carbon rod, and a slightly acid electrolyte which is in the form of a paste. The manganese dioxide acts as a depolariser to combine with the hydrogen as soon as it is formed, while the carbon rod acts as the current collector for the cathode.

The alkaline-manganese cell also has a zinc anode and a manganese-dioxide cathode, but in this case the electrolyte paste is alkaline. (Note the inconsistency in the way that this cell and the zinc-carbon cell are named.)

Another important cell design is the button cell – so called because of its shape, shown in fig. 12.4. It meets the demand for small cells for hearing aids, digital watches, and photographic equipment.

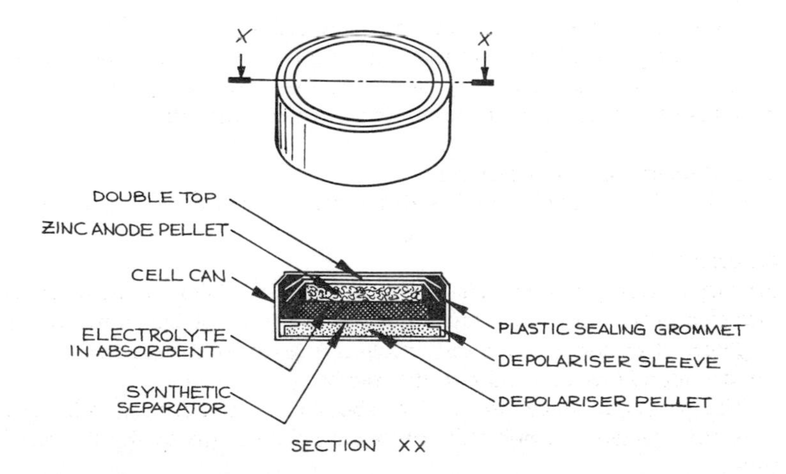

Fig. 12.4 The mercury (button) cell, which uses a zinc anode, a mercuric-oxide cathode, and a potassium-hydroxide electrolyte to provide a high-energy output.

Due to their small size, button cells require materials which will give a good performance. Here, mercuric oxide, although expensive, offers many advantages as a cathode. Another type of button cell uses a silver-oxide cathode. In both cases the anode is zinc.

Secondary cells

Secondary cells are rechargeable. The electrode materials are contained in porous plates, and this prevents serious distortion of either electrode during use and allows the cell to be recharged. This is done by pushing a small current through the cell in the direction opposite to the direction in which current was drawn from the cell when it was in use.

One type of secondary cell is the lead-acid accumulator used in motor cars to supply power for starting, ignition, and lighting. Since the electrodes are made of lead, the lead-acid cell has the disadvantage of being very heavy.

Modern accumulators consist of plates in the form of lead-antimony grids filled with special pastes under pressure, interleaved as shown in fig. 12.5. The positive plate is filled with red lead oxide, while the negative plate is filled with a lead oxide called litharge. On their first charge these plates become converted to lead peroxide and spongy lead respectively. The electrolyte is a dilute solution of sulphuric acid.

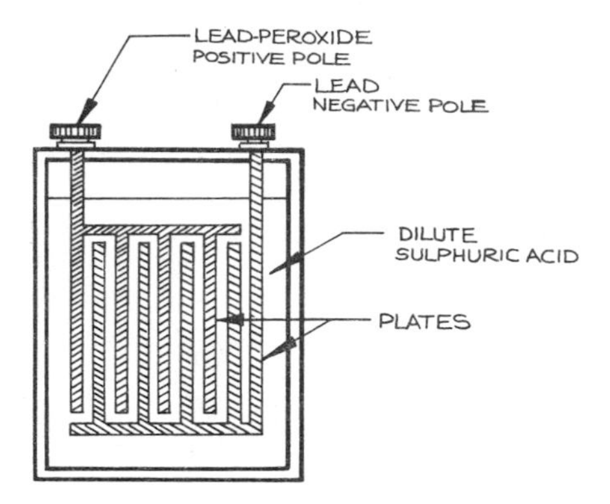

Fig. 12.5 The lead-acid cell – a secondary cell

As the cell, discharges, both plates slowly change to lead sulphate, and the acid becomes more dilute. The state of the cell may be determined by testing the acid using a bulb hydrometer. This measures the relative density of the acid, which should be 1.28 when the cell is fully charged. When the relative density of the acid has fallen to 1.12, the cell is regarded as fully discharged.

Lead-acid cells should be recharged regularly and not be left standing in a discharged condition for any length of time, otherwise the lead sulphate in the plates changes to a white crystalline form and is irrecoverable. This is referred to as 'sulphating'.

The lead–acid cell has a very low internal resisiance, due to the large surface areas of the plates and due to the fact that the plates are very close together. This means that the accumulator can supply large currents of the order of 30 A for short periods.

Other types of secondary (rechargeable) calls are referred to as alkaline cells. One is the nickel–iron or NiFe cell. Another is the nickel–cadmium or NiCd cell as used in calculators, radio control, and small power tools.

In both cells the positive plate is made of nickel hydroxide, while the electrolyte is a solution of potassium hydroxide (KOH). In the NiFe cell the negative plate is made of iron oxide, while the NiCd cell has a negative plate made of cadmium.

Alkaline cells are not subject to sulphating and can therefore be left in any state of charge without damage. Their disadvantages are higher cost than the corresponding lead–acid cell and the fact that they produce an e.m.f. of only 1.2 V as compared to the lead–acid e.m.f. of 2 V.

Experiment 12.1 To investigate the discharge characteristics of a lead–acid cell at various discharge rates.

The circuit is shown in fig. 12.6.

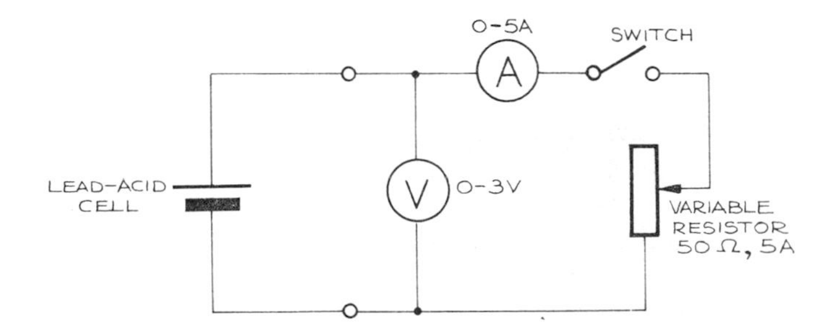

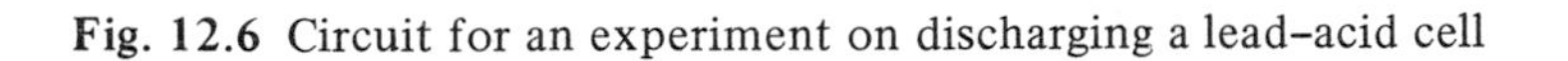
Fig. 12.6 Circuit for an experiment on discharging a lead–acid cell

Equipment

a) A lead–acid cell, fully charged
b) A 0 to 3 V d.c. voltmeter
c) A 0 to 5 A d.c. ammeter
d) A 50 Ω 5 A variable resistor (rheostat)

Notes on method

1 Connect the circuit as shown in fig. 12.6.
2 Adjust the load current to 1 A by variation of the variable resistor.
3 Read the terminal potential difference at hourly intervals until it has fallen to about 1.85 V.

4 Repeat using a fully charged lead–acid cell but with the load current set to 2.5 A.
5 Plot graphs of terminal p.d. against time.

Results
The readings obtained are shown in Table 12.1

Table 12.1 Terminal p.d. during discharge of a lead–acid cell

Time (hours)	Terminal p.d. (volts)	
	$I = 1$ A	$I = 2.5$ A
0	1.98	1.97
1	1.97	1.94
2	1.97	1.90
3	1.97	1.83
4	1.96	
5	1.96	
6	1.95	
7	1.90	
8	1.84	

The graphs are shown in fig. 12.7.

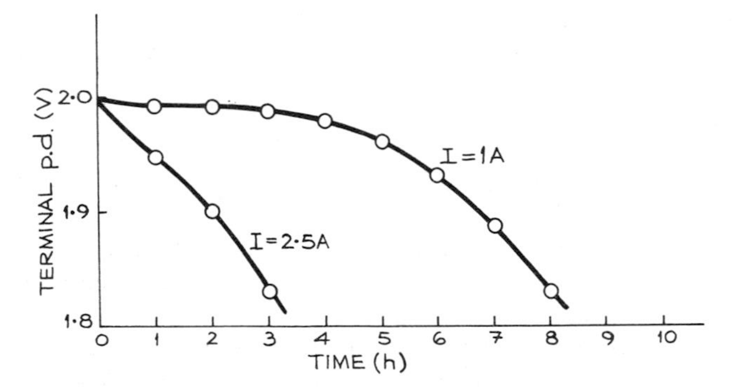

Fig. 12.7 Graph of terminal p.d. against time for a discharging lead–acid cell

Comments on results
The terminal p.d. decreases with time when taking current from the cell.

The rate of decrease depends on the load current, the rate of fall being more rapid with larger load currents.

Exercises on chapter 12

1 (a) State how an electrolyte conducts electricity. (b) State how pure water can be made into an electrolyte.

2 State the fundamental difference between a primary cell and a secondary cell and name one of each type.

3 A chromium anode is placed in an electrolyte together with an iron object, as shown in fig. 12.8. Explain (a) what is likely to happen some time after switch S has been closed and (b) why this happens.

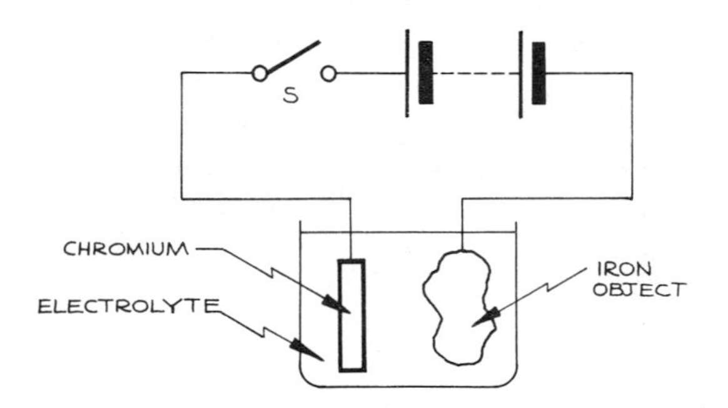

Fig. 12.8 Electrodes in a liquid electrolyte

4 (a) Explain the operation of a primary cell. (b) A student tries to make a cell by putting two copper electrodes in dilute sulphuric acid. It does not work. State what the student could change in order to make it into a simple cell.

5 Explain the process of electroplating with copper.

6 (a) Explain the action of a simple cell using electrodes of copper and zinc. (b) State two limitations of this cell. (c) State how these limitations are overcome in a zinc-carbon cell.

7 (a) Explain the difference between primary and secondary cells. (b) State the type of cell that is used in a motor car to operate the starter motor, and why this type is used.

8 (a) State two advantages and two disadvantages of a lead-acid cell. (b) State the advantages that alkaline cells have over lead-acid cells.

13 Electromagnetism

13.1 Introduction

Electromagnetism deals with the relationship between electric currents and magnetic fields. It plays a major part in our daily lives, as all the electrical devices that we take for granted - such as street lighting, domestic electricity supply, electric heating, etc. - could not be used without the availability of large quantities of electricity produced by generating stations. All of these generating stations use generators which depend for their operation on electromagnetism.

Before dealing with electromagnetism, let us consider magnetic fields.

13.2 Magnetic fields

A magnet affects the space surrounding it such that other magnets placed in this space experience forces. The space in which this occurs is known as a magnetic field. (The term 'field' is merely a way of saying that some effects are felt at a distance from the source of disturbance, rather like the heat felt around a hot body.)

The presence of a magnetic field surrounding a magnet may be demonstrated by sprinkling iron filings on to a sheet of thin card on top of the magnet. A pattern like that shown in fig. 13.1 is obtained.

Michael Faraday (1791-1867) referred to the apparent lines as lines of *magnetic flux* and, although the flux does not really exist as a number of separate lines, the concept is a very useful one since it provides a basis for explaining the various magnetic effects. For example, where the flux is more intense (i.e. where the magnetic effects are more pronounced) then the lines are closer together; and where the flux is weaker then the lines are more widely spaced apart.

The magnetic flux is strongest near the poles of the magnet, i.e. the points from which the magnetic flux seems to emanate. Thus the closeness of the lines roughly indicates the strength of the field.

The lines of magnetic flux form complete closed paths as shown in fig. 13.2. They appear as lines which start at one pole and form a closed link with the other pole. In areas some distance from the magnet, where the earth's magnetic field becomes noticeable in comparison with the field of the magnet, flux paths which do not close around the magnet join those of the earth's magnetic field such that continuity is still maintained.

Lines of magnetic flux do not intersect, although they may become very distorted. The lines behave rather like stretched elastic bands, in that they tend to try to shorten themselves.

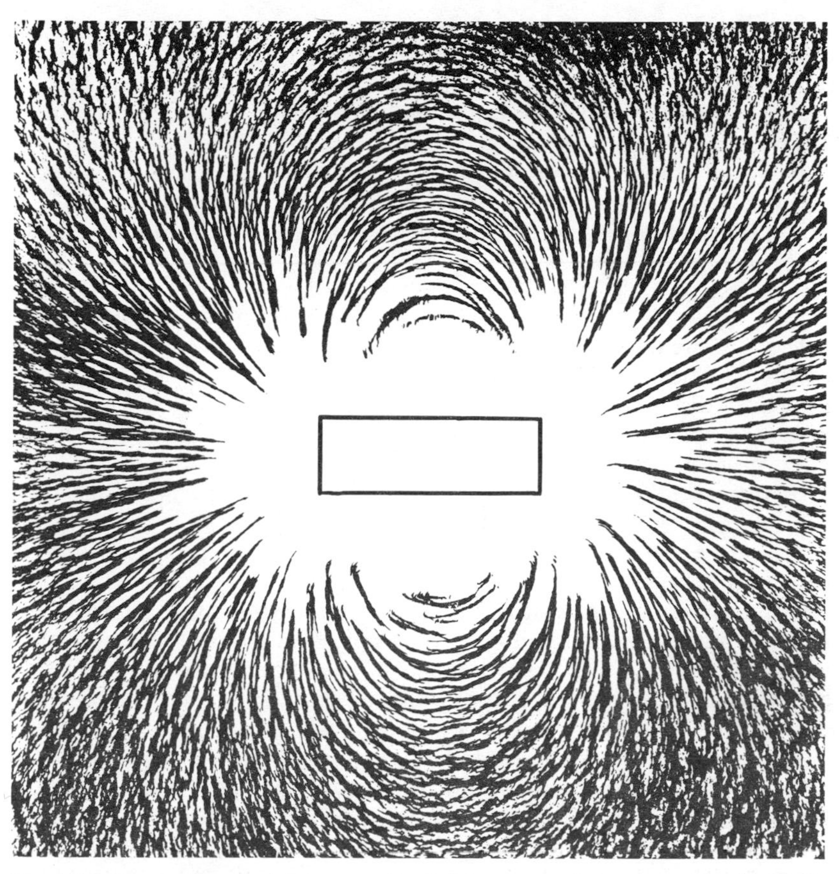

Fig. 13.1 The pattern of the magnetic field surrounding a magnet, obtained by using iron filings

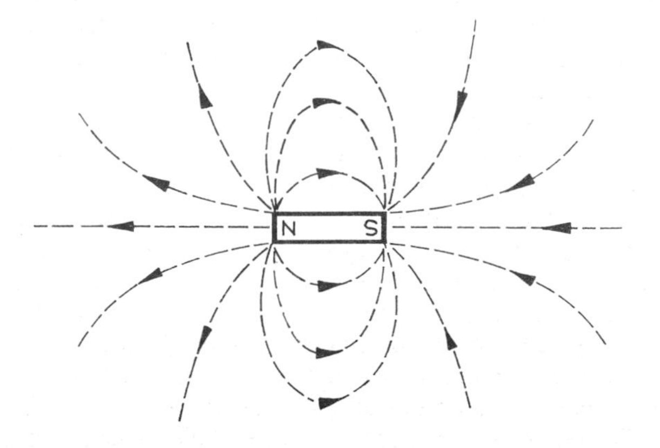

Fig. 13.2 Apparent lines of magnetic flux surrounding a magnet

Since fields are noticeable by their effects, we can define a direction in which these effects occur. The *direction* of each apparent line of flux is defined as being the direction in which a point north pole would travel (if such a thing could exist). This means that the direction of lines of magnetic flux is defined as being from a north pole to a south pole.

Point north poles do not exist but are merely a convenient concept - however, we can determine the direction using a compass as shown in fig. 13.3. The south-seeking pole of the compass points in the direction of the magnetic field.

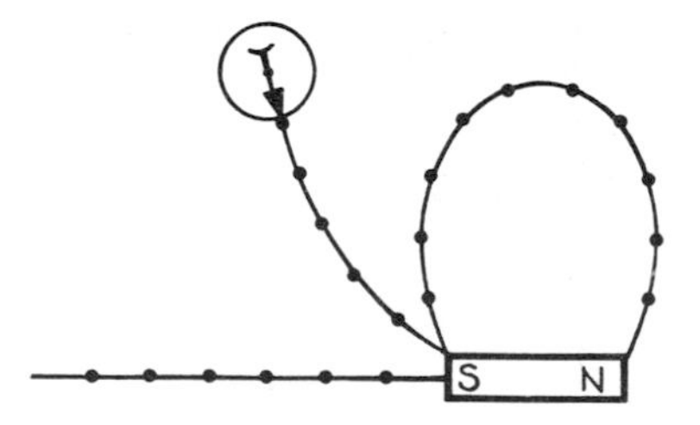

Fig. 13.3 A magnetic-flux pattern, obtained by using a plotting compass

13.3 Magnetic field due to an electric current

Now let us consider the phenomenon called electromagnetism, beginning with the magnetic effect of an electric current. Hans Christian Oersted (1777-1851) discovered in 1819 that, when an electric current flows through a conductor, it sets up a magnetic field around the conductor.

This may easily be demonstrated by winding a piece of wire around an iron or steel former, such as a nail, and connecting an electric cell across the wire as shown in fig. 13.4. If the nail is dipped into a container of iron

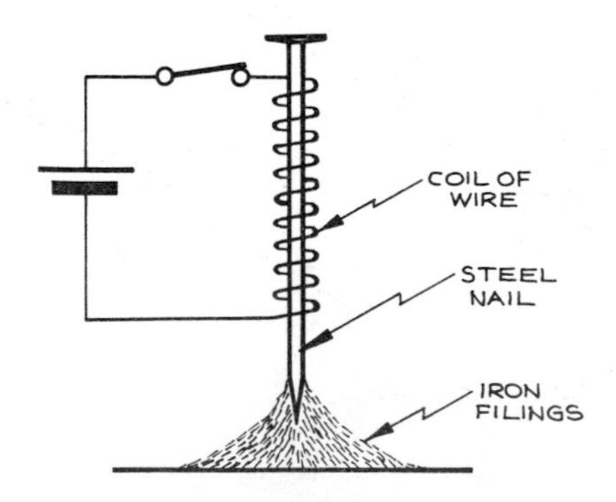

Fig. 13.4 Iron filings attracted to a steel nail, magnetised as a result of the current passing through a coil surrounding the nail

filings, some of the filings are attracted to the nail. The nail has become a magnet, magnetised by the electric current.

If we consider current passing through a single straight conductor, then the magnetic field surrounds the conductor like a continuous jacket and, using the concept of lines of flux, may be thought of as a series of concentric circles as shown in fig. 13.5.

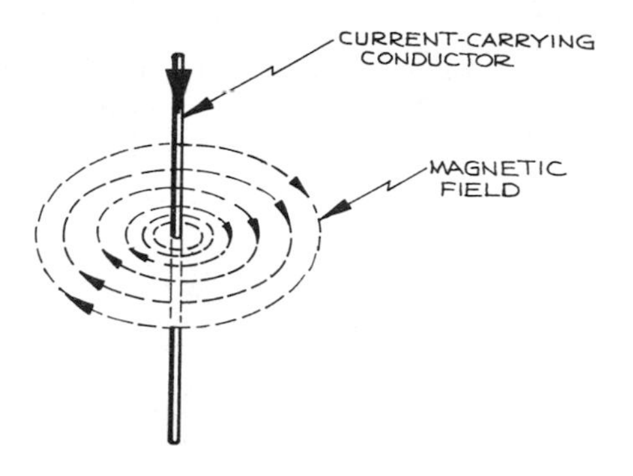

Fig. 13.5 The magnetic field surrounding a current-carrying conductor

Notice that the flux is most intense closest to the conductor, and this is represented by the lines being closer together in this region. Further away from the conductor, the flux becomes weaker, and therefore the lines are more widely spaced apart.

The direction of the lines of flux may be found by using the right-hand screw rule. This rule is shown in fig. 13.6, and states that, if the direction of the current in a conductor is considered to be the direction of travel of a screw, then the direction of the lines of flux will be the direction of rotation of the screw.

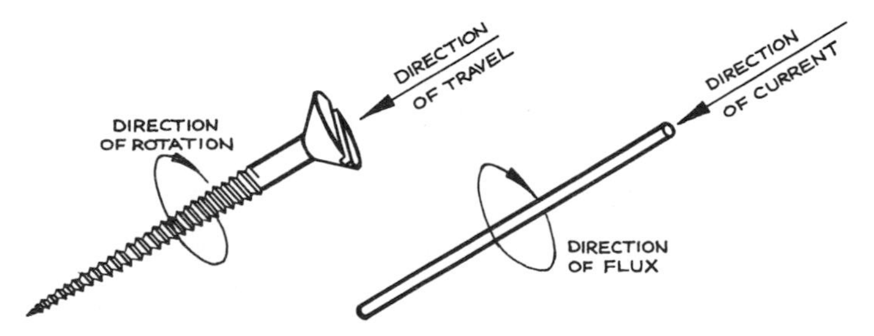

Fig. 13.6 The right-hand-screw rule

A diagrammatic convention is adopted to represent the direction of the electric current in a conductor which is end-on to the observer.

Figure 13.7(a) shows a conductor in which the current is flowing away from the observer. In fig. 13.7(b) the current is flowing towards the observer. The dot and the cross may be thought of as representing the point and flights of an arrow.

Fig. 13.7 Diagrammatic convention to represent the direction of the electric current in a conductor end-on to the observer: (a) current flowing away from the observer, (b) current flowing towards the observer

Example Use the right-hand screw rule to state the direction of the flux surrounding the conductors in figs 13.7(a) and (b).

The direction of the flux is (a) clockwise, (b) anticlockwise.

If a conductor is wound around a cylindrical former, it is then referred to as a coil or a *solenoid*. Passing a current through this solenoid produces a magnetic-flux pattern similar to that of a bar magnet, as shown in fig. 13.8.

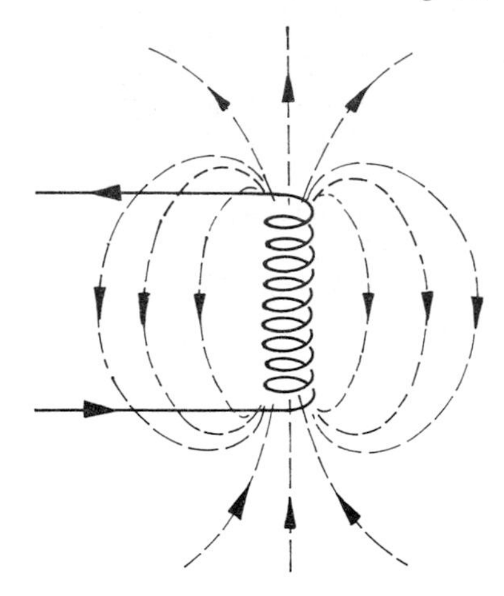

Fig. 13.8 The magnetic-flux pattern surrounding a current-carrying solenoid

The strength of the magnetic field around the solenoid depends on the number of turns of the coil and on the magnitude of the electric current. However, there is another factor which makes a tremendous difference to the strength of the magnetic field. Winding the coil on a soft-iron former rather than on, say, a wooden former can increase the strength of the magnetic field by as much as 1000 times. This is a very important effect. It is made use of in electromagnets, which consist of a soft-iron core surrounded by a coil of insulated wire.

If we wish to make good electromagnets, then we need to use formers (or cores) with good magnetic properties. This means using soft iron, and other materials will not do at all.

There are innumerable examples of the use of this electromagnetic effect, most of which use solenoids wound around soft-iron cores. Some examples are electromagnets for use in handling swarf etc; lifting magnets; transformers for converting alternating voltages from one level to another; relays which use a small electric current in a coil to close a set of contacts which are able to carry a much larger current; electric bells and buzzers; tape-recorders, in which the magnitude of the current fed into the tape-head coil controls the strength of the magnetic field which magnetises the tape; the telephone-receiver earpiece, in which a varying electric current passes through a solenoid, thus causing vibrations in a magnetic alloy diaphragm and so producing a sound; and many others.

A lifting electromagnet used for the separation of ferrous and non-ferrous metals

13.4 Electromagnetic induction

Electromagnetic induction is the basis of many electrical machines, such as the generator and the transformer.

The phenomenon of electromagnetic induction is the producing of an e.m.f. in a coil by a changing magnetic flux that links with the coil. Michael Faraday discovered this important effect in 1831 and it may be stated as follows: 'A coil situated in a magnetic field will have an e.m.f. induced in it if there is a *change* in the magnetic flux linking with the coil.' This is often referred to as Faraday's first law of electromagnetic induction.

It is important to realise that an e.m.f. is induced only while the flux is *changing* relative to the coil.

Consider the arrangement of fig. 13.9, in which a coil has a central-zero galvanometer connected across its ends. The galvanometer is an instrument for measuring voltage and will indicate any e.m.f. that is induced. If a bar magnet is pushed into the coil, a deflection of the galvanometer needle indicates that an e.m.f. is induced in the coil, as shown in fig. 13.9(a). This e.m.f. exists only while the magnet is being moved.

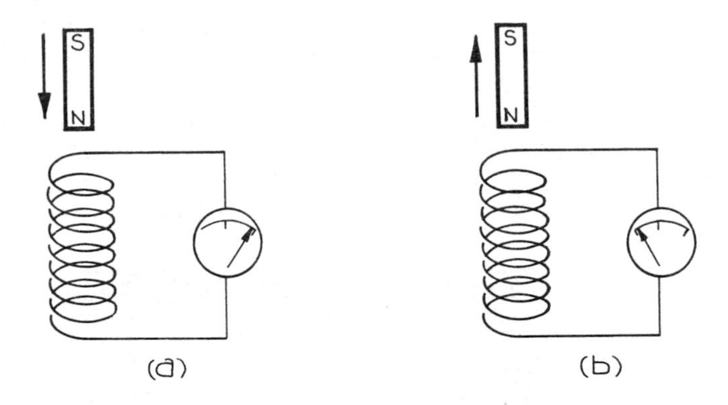

Fig. 13.9 Induced e.m.f. in a coil due to change of magnetic flux

If the magnet is now pulled out of the coil, an e.m.f. will again be induced - this time in the opposite direction, as shown in fig. 13.9(b).

The e.m.f. is in each case due to the change of the flux that links with the coil when the magnet is moved. No e.m.f. is induced when the flux stops changing, i.e. when the magnet stops moving.

Another means by which the magnetic flux may be changed, other than by movement, is to increase or decrease the magnetic field. Michael Faraday showed that an e.m.f. may be induced in a coil using this method, and in so doing he produced the forerunner of the modern transformer.

Faraday used a soft-iron ring wound with two separate insulated coils. One of the coils is called the secondary and has a galvanometer connected

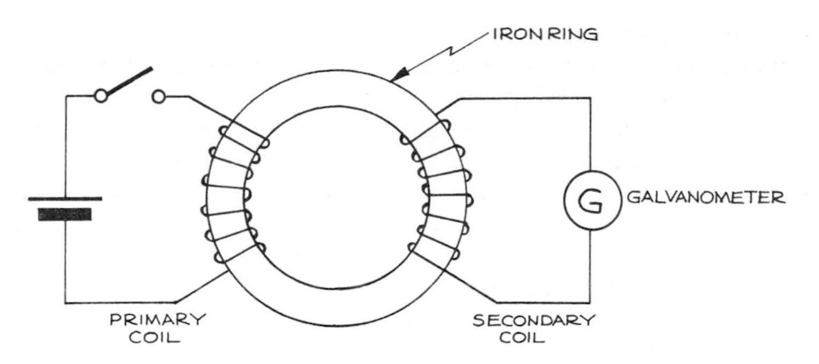

Fig. 13.10 A simplified representation of the Faraday ring

across its ends. The other coil is called the primary, and is connected via a switch to an electric cell as shown in fig. 13.10.

When the switch is closed, a current flows through the primary coil, thus producing a magnetic flux which exists mainly in the iron ring but which also links with the secondary coil. When the switch is opened, the current flow in the primary ceases, and the magnetic flux in the iron ring suddenly collapses. The growth and decay of the magnetic flux due to the closing and opening of the switch induces an e.m.f. in the secondary coil by electromagnetic induction, thus producing a deflection on the galvanometer.

By far the larger e.m.f. is produced when the switch is opened, since in this case the current and hence the magnetic flux is reduced to zero almost instantaneously.

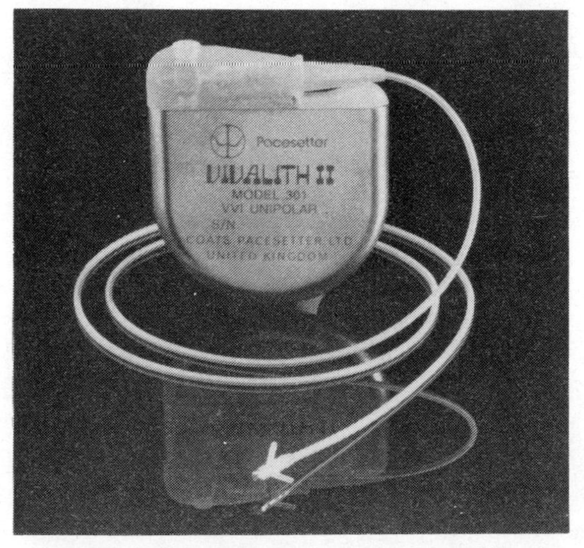

A heart pacemaker which makes use of the transformer effect to generate a pulse in a probe implanted in the heart and thus regulate heart rate

Applications of electromagnetic induction include the transformer, which is used to step up or down alternating voltages; the heart pacemaker, which makes use of the transformer effect to regulate heart beat; the induction furnace, in which heat is generated by induction of high-frequency currents in the material to be heated; the motor-car ignition system, which uses an induction coil, a contact-breaker, and a battery to produce the high voltage which is then applied to the spark-plugs; and many more.

Wolsung power station 800 MVA 345/26 kV generator transformer assembled for test supported on air-cushion equipment, showing the high-voltage side

Electric-induction melting – pouring iron from an Inductotherm 500 kg capacity medium-frequency furnace powered by a 250 kW 100 Hz 'VIP' power and control unit

13.5 The force on a current-carrying conductor in a magnetic field

When a current passes through a conductor situated in a magnetic field, a force acts on the conductor which tends to move it out of the field. Consider the following experiment.

Figure 13.11 shows a conductor suspended so that it is free to move and so that its lower end dips into a bath of mercury. The conductor may thus be connected, via a switch, to a battery, since mercury conducts an electric current. A horseshoe magnet surrounds the conductor as shown.

When the switch is closed, it is found that the conductor flicks out of the bowl of mercury. This is because a current flows in the conductor, and this current reacts with the magnetic field of the horseshoe magnet to produce a force which deflects the conductor out of the field.

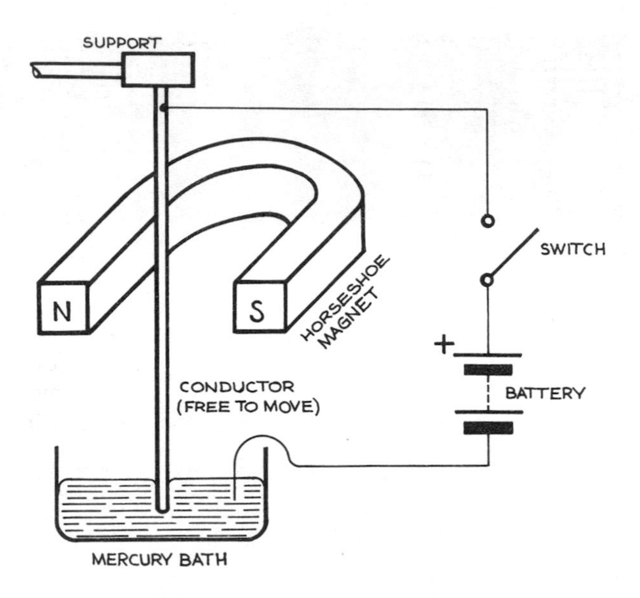

Fig. 13.11 The reaction of a current with a magnetic field to produce a force – the basis of the electric motor

This effect is the basis of the electric motor and also of the moving-coil instrument (for measurement of electric current) and the moving-coil loudspeaker.

One method of regarding this effect, which also provides a convenient method of finding the direction of the force, is to consider the force as being due to the interaction of two magnetic fields.

A current-carrying conductor, situated in a magnetic field, is shown in fig. 13.12. Considering each component separately, the magnetic field is shown in fig. 13.13(a) and the current-carrying conductor in fig. 13.13(b). Now we already know from section 13.3 that, when a current flows in a conductor, a magnetic field is set up around the conductor. The direction of the lines of magnetic flux are found by using the right-hand screw rule, and are as shown in fig. 13.13(b).

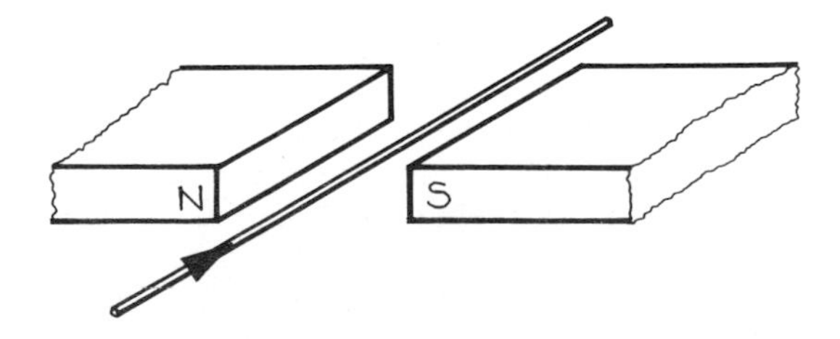

Fig. 13.12 A current-carrying conductor situated in a magnetic field

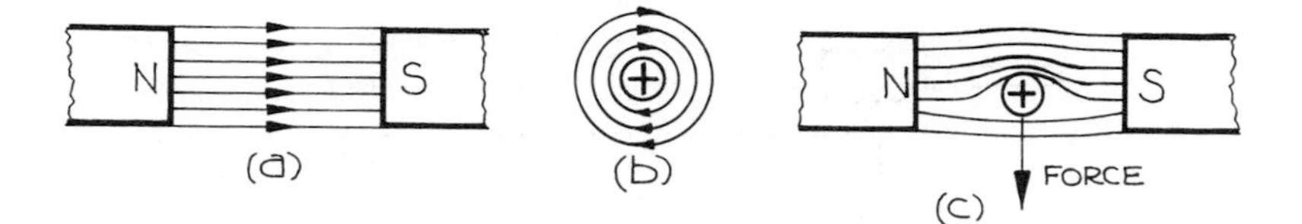

Fig. 13.13 Force on a current-carrying conductor situated in a magnetic field

If the conductor is now placed in the uniform field of fig. 13.13(a), then the two fields interact to form the pattern shown in fig. 13.13(c). Notice that above the conductor both of the flux directions are the same, so the total flux is increased; while below the conductor the two flux directions are in opposition and the total flux is reduced.

The flux pattern is stretched rather like a web of elastic bands with a bar pushed into it, and the result is that the conductor experiences a downward force as shown in fig. 13.13(c).

Notice that if the direction of the current were to be reversed then the force would also be reversed – i.e. it would act upwards.

Example A moving-coil loudspeaker is shown in fig. 13.14 and consists of a coil freely suspended over a soft-iron core. A tubular permanent magnet surrounds the coil, and the coil is connected by flexible leads to the terminals. Also connected to the movable coil is a paper cone. Explain the loudspeaker operation when an alternating e.m.f. is connected across the terminals.

When an alternating current flows through the coil, this current reacts with magnetic field to produce a vibratory motion on the coil former and thus

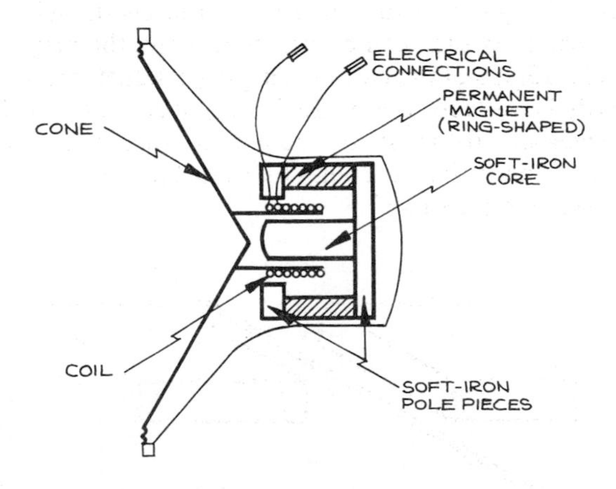

Fig. 13.14 A moving-coil loudspeaker

causes the paper cone to vibrate, so setting the surrounding air in vibration. If the alternating e.m.f. is a pure sine wave, a pure audible tone will be produced by the loudspeaker.

13.6 The moving-coil instrument

The permanent-magnet moving-coil instrument is used to measure electric current. It depends for its operation on the interaction between the current in a freely suspended coil and the field of a fixed permanent magnet.

It consists of a coil suspended in a magnetic field as shown in fig. 13.15. The current to be measured passes through the coil and reacts with the magnetic field to produce a turning-effort or torque.

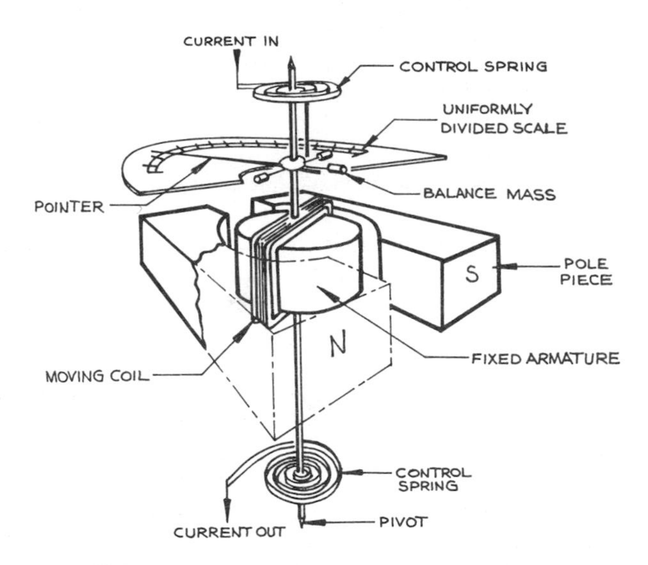

Fig. 13.15 A moving-coil instrument

In any current-measuring instrument, a restoring torque is required so that the instrument may be calibrated (i.e. so that the pointer does not swing right over to the end-stop whatever the current). This restoring torque is provided by control springs.

The moving coil has many turns, to provide a large torque, and is wound around a soft-iron core to provide a path of low magnetic reluctance in which the flux can readily exist. (Reluctance in magnetic circuits is similar to resistance in electric circuits.) The coil is balanced on pivots at each end, and a pointer connected to the coil moves across a scale as shown.

The current is fed to the coil via the control springs. The reading is given on the dial when the restoring torque set up by the control springs is equal to the deflecting torque of the coil.

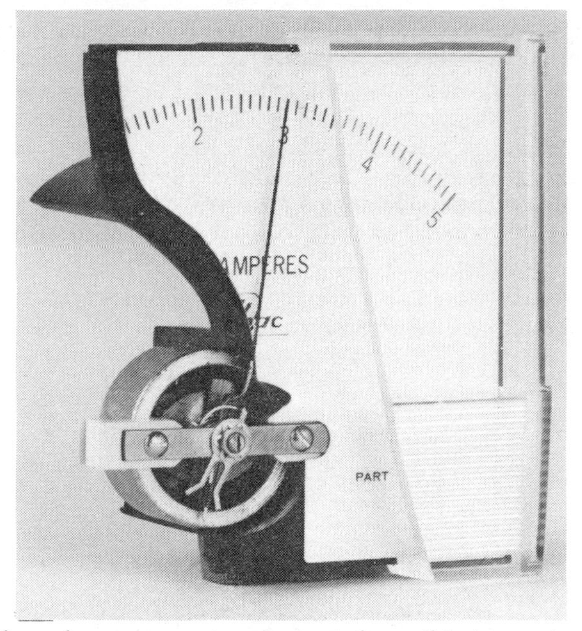

A section through a permanent-magnet moving-coil instrument used as an ammeter

The pole-pieces of the permanent magnet are shaped so that the coil moves through a uniform magnetic flux over its entire movement. This is important since it ensures that the instrument has a linear scale.

Damping of the movement must be provided in any moving-pointer instrument, to reduce oscillations and prevent overshoot of the pointer. In the case of the moving-coil instrument, this damping is provided by winding the coil on an aluminium former. When the coil moves through the magnetic field, small circulating currents (referred to as 'eddy currents') flow in the aluminium former. These currents react with the magnetic field to produce a restraining torque while the coil is in motion and thus damp out oscillations. Instrument damping is generally arranged to allow just one small overshoot, to show that the pointer is not sticking.

One disadvantage of the moving-coil instrument is that it will not measure alternating current, since this would attempt to move the pointer backwards and forwards at the frequency of the alternating current and would effectively result in no motion at all.

Example Figure 13.16 shows the end-on view of two coil sides of a coil in a moving-coil instrument. Draw the lines of magnetic flux due to the magnet and those due to the current in the coil sides, and hence show that the motion of the coil will be anticlockwise.

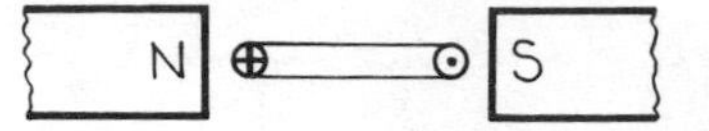

Fig. 13.16

The magnetic flux patterns are shown separately in figs 13.17(a) and (b). They interact as shown in fig. 13.17(c) to produce an anticlockwise torque on the coil.

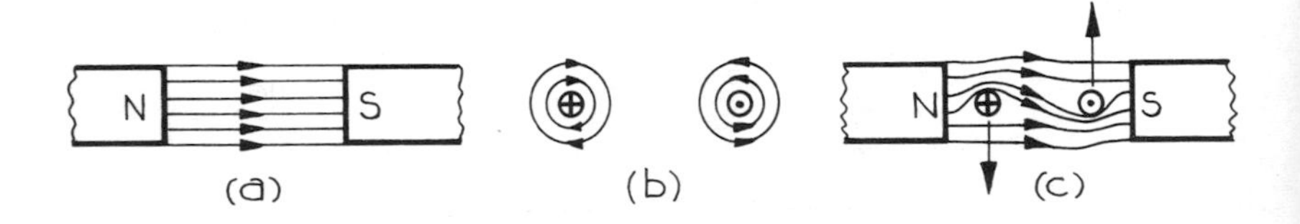

Fig. 13.17

Exercises on chapter 13

1 For the relay circuit shown in fig. 13.18, explain what happens when the switch S is closed.

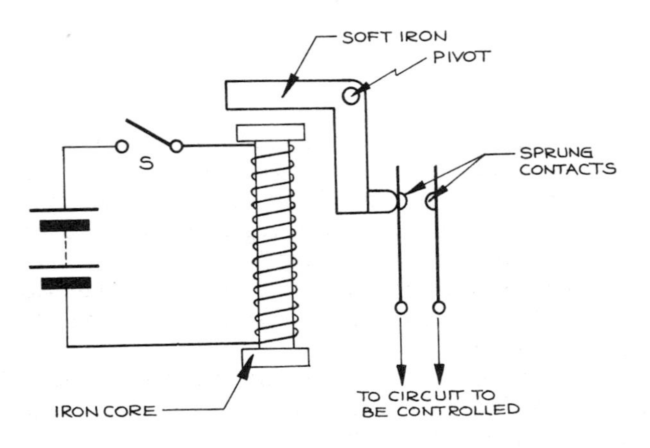

Fig. 13.18 A relay circuit

2 Sketch lines of force to indicate the form of the magnetic field around the bar magnet shown in fig. 13.19. The earth's magnetic field is such that north lies to the top of the paper.

3 Explain what happens when the switch is closed and then opened again in the circuit shown in fig. 13.20. Give reasons for your answer. Explain how this system could be made automatic so as to produce an electric bell.

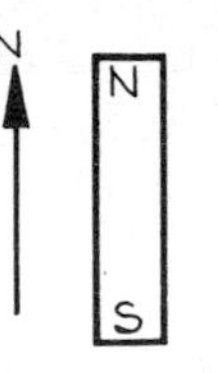

Fig. 13.19 A bar magnet

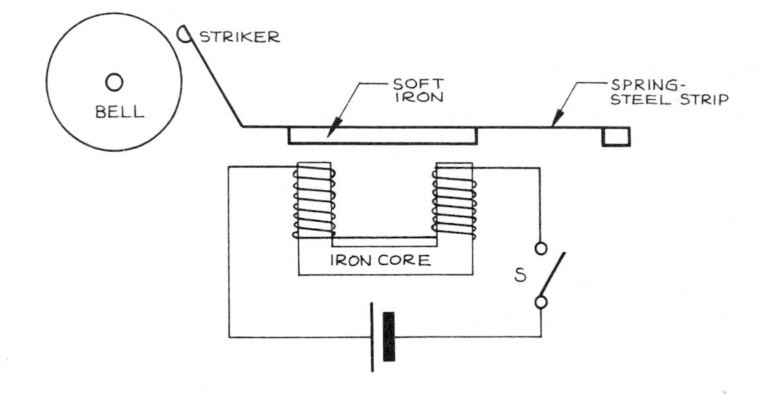

Fig. 13.20 A bell striker

4 (a) Sketch the magnetic-field pattern surrounding a bar magnct. (b) Sketch the magnetic-field pattern surrounding a current-carrying solenoid.
5 Draw a diagram of the operational parts of a moving-coil meter.
6 (a) State three applications of electromagnetism. (b) State three applications of electromagnetic induction.
7 Describe how the magnetic field of a bar magnet could be shown using a bar magnet, a sheet of cardboard, and iron filings. Make a sketch of the pattern that would result.
8 Make a sketch to show the magnetic-field pattern produced by currents flowing in two parallel conductors if the currents are (a) in the same direction, (b) in opposite directions.
9 Two coils are placed side by side. If a d.c. supply e.m.f. is suddenly switched across one coil, state what happens in the other coil and explain why.
10 Figure 13.21 shows two conductors of a d.c. generator armature. Assuming that the loop rotates in a clockwise direction, as indicated, show the direction of the e.m.f. in each conductor.
11 Figures 13.22(a) and (b) show a current-carrying coil situated in a magnetic field between two permanent magnets. Draw three sketches to show (a) the magnetic field produced by the permanent magnets only,

(b) the magnetic field produced by the current-carrying coil only, (c) the combined magnetic field. Indicate on sketch (c) the direction in which the coil will try to move.

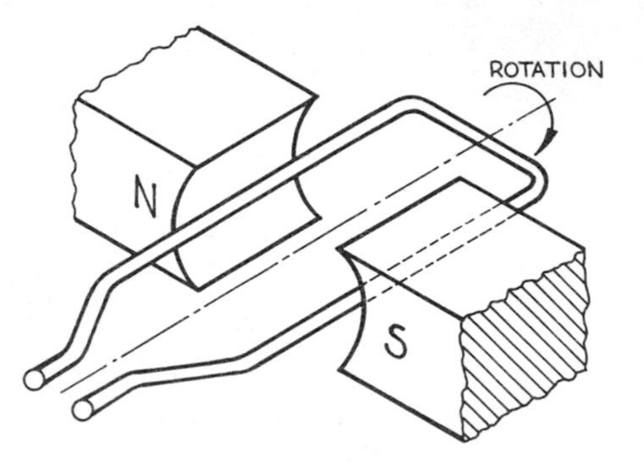

Fig. 13.21 Two conductors of a d.c. generator armature

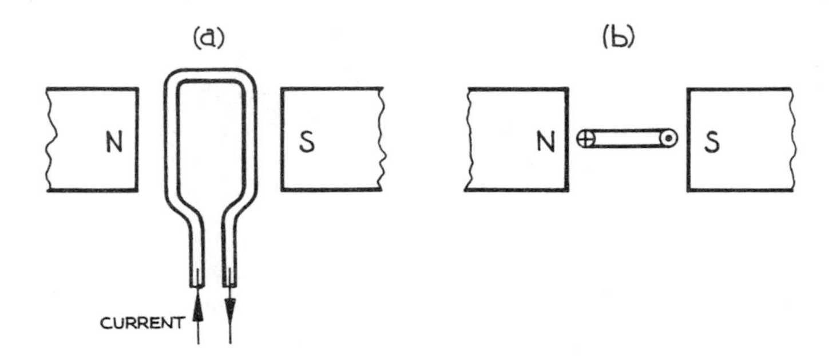

Fig. 13.22 A current-carrying coil situated in a magnetic field

Answers to numerical exercises

Chapter 1

1 (a) 63 kPa, (b) 50 nF, (c) 487 Gg, (d) 27.54 mm
2 (a) 92.5×10^{-12} F, (b) 7.568×10^{6} m, (c) 56.3×10^{12} m, (d) 760×10^{6} kg
3 (a) 24.414×10^{-9} m, (b) 7×10^{3} W, (c) 5×10^{-6} m, (d) 8×10^{21} Ω

Chapter 3

1 192 N
2 5 mm
3 300 N
4 (a) 60 N, (b) 3.67 mm
5 0.0173 mm
6 (a) 1 mm, (b) 3 mm, (c) 3.2 mm
7 (a) (i) 1.28 mm, (ii) 2.74 mm, (iii) 4.18 mm; (b) (i) 25 N, (ii) 75 N, (iii) 212 N
8 3.4 kN, 0.0095 mm, 3.275 kN
9 9.7 mm, 10.1 mm, 10.2 mm
10 0.124 mm
12 0.8 mm
13 2.45 mm
14 10 mm

Chapter 4

3 150 N
4 (a) 40 N m, (b) 37.5 N m, (c) 5 N m, (d) 10 N m
5 46.67 N
6 x_A = 4.44 mm, x_C = 1.11 mm
9 182.35 mm from AB, 100 mm from BC
10 400 N
11 (a) 0.6 N m, (b) 20.78 N m, (c) 30.64 N m, (d) 6.93 N m, (e) 30.07 N m
12 163.3 N
13 4414.5 N
14 44.6 N

Chapter 5

1 3.6 kN
2 (a) 2 N/mm^2, (b) 600 N
3 (a) 0.5 N/mm^2, (b) 500×10^{3} Pa, (c) 5 bar
4 9.12 kN
6 (a) 3.01 bar, (b) 301×10^{3} Pa
7 3.1009 MPa
8 (a) 1.014 bar, (b) 101.4×10^{3} Pa
9 (a) 730.8 mm, (b) 750.8 mm, (b) 750.28 mm, (c) 743.54 mm
10 6.67 kPa
11 1.472 MPa
12 157.55 kPa, 130.57 kPa
14 (a) 697.9 Pa, (b) 101.1979 kPa

Chapter 6

1 (a) 10 m/s, (b) 13.3 m/s, (c) 4.17 m/s, (d) 20 m/s, (e) 27.78 m/s, (f) 4 m/s, (g) 0.6 m/s
2 (a) 10.8 km/h, (b) 86.4 km/h, (c) 4.8 km/h, (d) 0.864 km/h
3 (a) 350 m, (b) 271.5 m; 3.044 m/s, 2.36 m/s
4 2725 km, 2524.1 km
5 72 km
6 $v_A = 0, v_B = 1.037$ m/s, 8.33 m/s for both
7 (a) 5.33 m/s^2; (b) (i) 48 m/s, (ii) 80 m/s
8 (a) 60 m/s, (b) 68 m/s
9 (a) 144 m; (b) 0.0033 m/s^2, 0.0067 m/s^2; (c) 0.0033 m/s^2
10 19.62 m, 19.62 m/s
11 (a) 985 m, (b) 1.2 m/s^2, (c) 0.5 m/s^2
12 2.5 m/s^2, 6.25 m/s^2, 418.75 m
15 2.473 s, 24.26 m/s
16 (a) 3 h 40 min, (b) 92.18 km/h or 25.6 m/s, (c) 74.73 km/h or 20.76 m/s

Chapter 7

1 289 m/s
3 1.5 s
4 430 000 GHz
6 0.335 m, 1.36 m
7 3.32 m to 3.16 m
8 300 GHz to 300 000 GHz

Chapter 8

1 750 kJ
2 200 N
3 5 m
4 275 J
5 7500 J
6 14.4 kJ
9 95%
10 200 kJ
11 62.5%
12 36 kW
13 3924 W
14 120 kg/h
15 10.13%
16 11.15 J, 0.0155 W

Chapter 9

1 83.6 kJ
2 1.504 kg
4 6.28 MJ
7 19.451 kJ
8 1.914 MJ
9 14.408 MJ
11 6628 J
12 336.96 kJ, 20 min

Chapter 10

2 (a) 45 V, (b) 3 A
3 360 Ω
4 20 Ω
5 0.4 A
6 3 Ω, 5 V
7 (a) 2 A, (b) 15 Ω
8 17 μΩm
14 0.16 A
15 6.8 kΩ
16 3.69 A
17 960 Ω
18 2 V
19 0.39 A
20 14 Ω
22 (a) 28.8 Ω, (b) 25 A
23 0.45 A
24 (a) 120 V, (b) 231 Ω, (c) 2.08 A
25 5.7 A
26 260 Ω
27 (a) 8 Ω, (b) 30 A

Chapter 11

1 20%
2 2.5 kW
3 10 A
4 0.1 W
5 8.64 MJ
6 10 V
7 24.98 MJ
8 (a) 0.2 kWh, (b) 1 p
9 4.5 kWh, 12.96 MJ
10 5 p
11 20.83 Ω, 12 A, 1.92 kW
12 £3, 12.5 A

Index